Matthew J. Liberatore (Ed.)

Selection and Evaluation of Advanced Manufacturing Technologies

With Contributions by
M. R. Blinkhorn, J. Borden, I.-J. Chen, C.-H. Chung
N. Danila, T. Dimnik, R. C. Dorf, L. Ferriera
B. Gold, J. Guarino, M. J. Liberatore, T. W. Lin
V. G. Lintner, A. Mehrez, F. Partovi, M. J. Pokorny
P. L. Primrose, J. M. Reeve, J. Sandholzer
W. G. Sullivan, D. Wilemon, M. M. Woods

With 23 Figures

Springer-Verlag
Berlin Heidelberg New York
London Paris Tokyo
Hong Kong Barcelona

Professor Matthew J. Liberatore
Department of Management
College of Commerce and Finance
Villanova University, Villanova
Pennsylvania 19085-1678, USA

Softcover reprint of the hardcover 1st edition 1990

ISBN-13: 978-3-642-95623-2 e-ISBN-13: 978-3-642-95621-8

DOI: 10.1007/ 978-3-642-95621-8

CONTENTS

INTRODUCTION

Matthew J. Liberatore
Department of Management
Villanova University
Villanova, PA 19085

1. BACKGROUND

The weakening competitive position of many segments of U.S. manufacturing has been analyzed, debated and discussed in corporate boardrooms, academic journals and the popular literature. One result has been a renewed commitment toward improving productivity and quality in the workplace. The drive to reduce manufacturing-related costs, while meeting ever-changing customer needs, has led many firms to consider more automated and flexible manufacturing systems. The extent to which these new technologies can support business goals in productivity, quality and flexibility is an especially important issue for manufacturing firms in the U.S. and other Western nations. Problems have arisen in developing performance measures and evaluation criteria which reflect the full range of costs and benefits associated with these technologies. Some would argue that managerial policies and attitudes, and not the shortcomings of the equipment or manufacturing processes, are the major impediments to implementation (Kaplan 1984).

Advanced manufacturing technologies (AMT), the subject matter of this volume, can be considered to span a continuum from stand-alone to linked to integrated, as discussed in Meredith and Suresh (1986). For example, robotics and computer numerically controlled (NC) machine tools are often in the stand-alone category, and can be linked into cells such as in group technology (GT) or flexible manufacturing systems (FMS). Other examples of linked systems include computer aided design (CAD) with computer-aided process planning (CAPP), automatic storage/retrieval systems (AS/RS) and manufacturing resource planning (MRPII). When the planning, design, manufacturing, materials handling and support systems (such as purchasing, cost accounting, and so forth) are all linked together through computer control, the factory is said to achieve full integration, called computer integrated manufacturing (CIM).

The strategic nature of investment in AMT must be recognized by manufacturing firms. For example, if a firm desires to become a world-class manufacturer, what level and type of investment in AMT is required? Traditional notions such as fixed versus variable cost and economies of scale are giving way to new notions such as activity-based costing and economies of scope. As a result, many firms have found it difficult to economically "justify" AMT investment using the standard methods of accounting and financial analysis. New multiple criteria methods are needed to fully reflect all the costs and benefits associated with AMT. These non-financial factors may include inventory (working capital, floor space), quality improvement, flexibility (economies of scope, product life cycle, product changes over time, backup capacity), throughput, lead time and organizational learning (Kaplan 1986). Including these factors into the evaluation process presents important challenges to managers and management scientists alike.

The systems used to control manufacturing also require new financial and operating measures of performance. These measures should be consistent with the criteria used in the initial investment evaluation. That is, performance measures should reflect the strategic goals of the firm, such as the requirements for long-term survival and/or growth. Current management accounting systems are sometimes inadequate because of their short-term emphasis and focus on the allocation of fixed costs. Some of the problems and issues include increasing overhead rates, shrinking base of direct labor over which to allocate costs, and increasing fixed costs and decreasing variable costs (Kaplan 1984). The need for better performance measures and evaluation criteria has been summarized by C. Jackson Grayson, Jr. and Carla O'Dell (1988) as follows:

> Trying to make twentieth century decisions based on eighteenth-century accounting principles is like driving a car with the emergency brake on. You can do it, but there will be a lot of smoke and screeching.

2. PURPOSE

The purpose of this book is to provide a unified treatment of the current state of knowledge concerning the evaluation and selection of AMT. A multi-functional perspective is presented, and contributions from accounting, management science, technological innovation, strategic planning and organizational behavior are included so that a broad-based understanding of this topic can be achieved. This book is managerial in its orientation. Extensive mathematical treatments are minimized, and sufficient references are provided wherever needed. New selection and evaluation concepts and methods are presented in conjunction with case studies or application discussions.

This book is addressed to three types of readers:

(1) managers and consultants who evaluate AMT investment proposals and/or recommend changes in the accounting systems used to track AMT performance;

(2) academics and professionals from management science, strategic planning, management accounting, and other related disciplines who are interested in developing better methods and systems for AMT evaluation and selection; and

(3) management scientists and other researchers working in the field of multiple criteria decision making (MCDM), since AMT evaluation and selection is an important and emerging MCDM application.

3. SUMMARY OF CONTENTS

This book is divided into three sections. In the first section, AMT evaluation and selection is viewed from the technological innovation perspective. The first chapter by Gold reviews the limitations of current practices for evaluating technological innovations in manufacturing, and offers suggestions for revising them based on the results of field studies in the U.S. and elsewhere. Guarino and Wilemon then consider how a manager's own objectives can influence the evaluation process. The next two chapters consider the adoption and diffusion patterns of specific forms of AMT, and relate their findings to a consideration of AMT benefits and evaluation criteria. Specifically, Dorf and Sandholzer study the use of programmable controllers in the U.S., while Pokorny, Lintner, Woods, and Blinkhorn investigate CADCAM usage in the U.K. mechanical engineering industry.

Part II contains four papers which consider specific approaches for AMT selection and evaluation. Reeve and Sullivan review a variety of methods for evaluating interrelated AMT investments. Chapters by Partovi and Danila present selection processes which are linked to the strategic plan of the organization. Primrose describes how standard financial analysis approaches can be applied to AMT evaluation through the full consideration of benefit and cost factors. All three authors provide case studies to support the application of their evaluation methods and processes.

Performance measurement and the management accounting system is the subject in Part III. This section contains five chapters and begins with Borden's review of the performance measurement and product costing literature. Ferreira and Lin then discuss some specific problems of performance measurement and cost control and conclude with a number of specific recommendations. Dimnik investigates the same issues through an empirical study which examined the influence of the management accounting system on the adoption of flexible automation. The section concludes with two articles on the measurement of flexibility, an important performance characteristic. Chung and Chen explore the relationships between flexibility, productivity, and competitiveness, while Mehrez suggests an economic measure of flexibility based on value of information concepts.

This book was developed on the basis of a call for papers. After a review of abstracts and chapter proposals, approximately 25 manuscripts were solicited and subjected to a peer review process. The 13 revised papers that are published in this volume represent those manuscripts that were accepted by the referees and the editor.

4. ACKNOWLEDGMENTS

The editor gratefully acknowledges the contributions of the chapter authors and the many reviewers that participated in this project. To Dean Alvin A. Clay and the staff of the College of Commerce and Finance, Villanova University, I owe thanks for continued support of this project. Special thanks are given to my graduate assistants, Joanne Conrad and Maryann Guro, for their help in organizing and managing the reviewing process and preparing the final manuscript. In addition, I wish to thank Dr. Werner A. Muller, Economics Editor for Springer-Verlag, for his encouragement and support. I thank my wife Mary Jane, and my children, Kathryn and Michelle, for their encouragement, cooperation and patience during the period over which this book was prepared.

5. REFERENCES

Grayson, C. Jackson, Jr. and Carla O'Dell. (1988). **American Business, A Two Minute Warning: Ten Changes Managers Must Make to Survive in the 21st Century.** New York: Free Press. London: Collier Macmillan.

Kaplan, Robert S. (1984). "Yesterday's Accounting Undermines Production," **Harvard Business Review**, July-August, 95-101.

Kaplan, Robert S. (1986). "Must CIM Be Justified By Faith Alone?" **Harvard Business Review**, March-April, 87-95.

Meredith, Jack R. and Nallan C. Suresh. (1986). "Justification Techniques for Advanced Manufacturing Technologies." **International Journal of Production Research, 24,** 1043-1057.

SOME NEEDED ADVANCES IN THE EVALUATION OF TECHNOLOGICAL ADVANCES

Bela Gold
Fletcher Jones Professor of Technology and Management
The Claremont Graduate School, Claremont, CA, USA 91711-6184

ABSTRACT

After reviewing some shortcomings of managerial approaches to improving technological capabilities, the discussion focuses on: establishing more effective targets for improvements; more useful guides for evaluating prospective innovations; and sounder bases for appraising and improving their results after installation.

1. INTRODUCTION

Achieving technological competitiveness is not a sufficient condition to ensure market competitiveness, but it is a necessary condition. Even advantages in marketing and in finance are bound to prove inadequate over time if products continue to be inferior in capabilities as well as more costly. Yet this requirement is only belatedly becoming recognized because its dominant role in the early emergence of American superiority in industry after industry came to be superseded in turn by increasing pressures to market our rapidly expanding output potentials and then by the need to respond to the ensuing demands for greater financial resources.

But the resulting prolonged under-emphasis on advancing the technological capabilities of our long-established major industries has necessitated re-establishing the technological foundations of their continuing competitiveness. Indeed, the increasingly dominant role of marketing and finance specialists in industrial leadership at large has tended to encourage inadequate commitments to seeking major advances in product development and productive efficiency even in our newer major industries, thereby seriously threatening their continued competitiveness as well.

2. MANAGERIAL APPROACHES TO IMPROVING TECHNOLOGICAL CAPABILITIES

The technological competitiveness of firms and industries is determined not by the rate at which significant innovations are developed, but by the extent to which they are applied to commercial operations. More than 35 years of analyzing the problems of improving industrial productivity have convinced me that implementing major advances in technology represents a far more difficult and far-reaching challenge to management than is generally appreciated. The key reason for this is the failure to recognize that basic technologies are built not only into the production machinery, but also into:

(1) the expertise of the technical personnel;

(2) the structure and operation of the production system;

(3) the economically feasible range of changes in product mix;

(4) the skills and organization of labor; and

(5) even the very criteria used to evaluate new capital goods proposals.

These represent powerful and mutually reinforcing commitments to preserving existing production and organizational arrangements, except for small, gradual and localized changes. The influence of such deep-rooted commitments was perceptively captured by the poetic lines:

> "With their eyes firmly fixed upon the past,
> They backed reluctantly into the future".

It would, of course, be undiplomatic in modern industry to voice immediate opposition to proposals for major innovations. But such resistance can be exerted quite effectively through unencouraging evaluations of proposals by the very specialists on whom managements depend for expert judgments, as well as by highlighting the uncertainties and difficulties likely to be encountered in applying and utilizing new technologies. Moreover, even when the need for improvement has come to be accepted, the targets are often set too low.

The development of programs to achieve major advances in technological competitiveness seems to require basic changes in traditional approaches:

(1) to generating promising new proposals;

(2) to evaluating proposals for adopting available technological innovations; and

(3) to evaluating the effectiveness with which resulting potentials have been harnessed.

Accordingly, the following discussion will review the limitations of widespread practices in each of these areas before offering suggestions for revising them on the basis of our field studies in a variety of industries in the U.S. and abroad.

3. GENERATING PROPOSALS FOR TECHNOLOGICAL INNOVATIONS

3.1 Weaknesses of Traditional Approaches

In most firms, proposals for technological innovations are expected to emerge from, or at least to be approved by, the operating sectors most likely to be affected by them. However, in the cases of robotics and, to an even greater extent, computer-aided manufacturing, for example, most operating staffs lack sufficient expertise to evaluate either specific applications or the broader alternative systems which are available for adoption. Moreover, the readiness of current operating executives and engineers to add personnel with such specialized capabilities may be inhibited by the prospect that increasing reliance on such new technologies would tend to increase the newcomers' opportunities for advancement at the expense of those lacking their expertise.

Managements also frequently act as though the most promising innovational proposals are bound to keep flowing because of an unrestrainable urge of managers and technical specialists to maximize improvements. In fact, however, analysis suggests that both the volume and the magnitude of innovational proposals are closely responsive to the attitudes of top managements, as reflected not by their exhortations but by their decisions about the proposals which have been submitted. Thus, a high ratio of rejections of new kinds of innovational proposals is bound to discourage additional submissions; and evidences that management favors only modest scale or quick repayment proposals tend to evoke conforming adjustments in the characteristics of the proposals which are submitted. It may be of interest to contrast the steady and impressive progress of computerization in the integrated steel industry of Japan, with its top-down guiding strategy, with the intermittent and more limited gains resulting from reliance on the earlier but essentially bottom-up efforts in the American integrated mills (Gold, 1979).

3.2 Establishing Targets for Needed Improvements

One of the greatest obstacles to increasing the competitiveness of many manufacturing firms has been the inadequate awareness of top managements of their own responsibilities for coping with looming threats due to technological lags (Gold, 1985 and Gold, 1983). This has been attributable partly to their understandable preoccupation with the financial aspects of performance. But is has also been due to technical advisers who have either lacked an adequate grasp of new technological advances, or who have been unwilling to make forceful presentations of unwelcome urgencies.

It must be recognized, however, that major technological improvements usually involve substantial investments and risks in addition to taking extended periods before their full potentials are likely to be realized. Therefore, development of an effective technological improvement program should be based on a careful diagnosis of the key pressures and needs which lie ahead - although action-oriented officials are often impatient with such analytical efforts.

Such needed diagnosis should center around three tasks:

(1) Identifying not only current but also prospective threats over at least the next 3-5 years, in order to take account

of the fact that competitors, too, are likely to have technological improvement programs underway. This involves serious efforts to appraise the firm's relative advantages and disadvantages in each product line and in each market as a basis for estimating the relative rewards likely to result from increased resource commitments to these various sectors;

(2) Appraising the risks, costs and problems involved in making alternative combinations of commitments, including the adoption of already available technologies as well as the initiation of programs seeking internally developed advances - in comparison with assessments of probable concomitant improvement efforts by competitors;

(3) Exploring the adjustments in finance, marketing and even personnel capabilities and organizational arrangements which are likely to be necessary in order to harness resulting technological potentials.

In order to ensure progress towards such improvement goals, it might be useful to begin by establishing a special group, closely tied to top management, to undertake the current and longer term evaluations suggested above. This group could then build on the findings as a basis for proposing preliminary improvement targets, and to draw on all kinds of internal and external expertise to evaluate alternative innovations capable of furthering progress toward such goals.

Key questions might include: which advances should be licensed or bought from the outside and which should we seek to develop internally? How many of needed advances can be achieved by modifying or upgrading existing facilities and products as compared with those gains available only through shifting to new facilities and products in the next two or three years, and over the next five years? Issues like these require consideration of the proportion of current and prospective capital outlays which should be allotted to "strategic investments", meaning those to be evaluated on the basis of expected contributions to longer term competitiveness and profitability.

Decisions about such matters are often rooted in basic philosophical commitments rather than in specified analytical procedures. For example, during the course of one of our field studies, we heard a director of manufacturing technology support a proposal for the acquisition of major new facilities and equipment with a comprehensive array of estimates of investment requirements, operating costs and revenues over the next six years, only to have them abruptly dismissed by the capital allocations official as "nothing more than unreliable guesses which could not justify the large investments involved." "But", the latter then added, "tell me why the company needs to make such an investment in order to safeguard its future competitiveness". In response to our query about the basis for taking such an unusual position, he told us that, in order to maintain its outstanding position in the industry, this company consistently allocates 15-20% of its capital expenditures to projects which are too new to permit persuasive analysis of prospective returns, but which seem of such great future promise as to warrant undertaking significant investments in pioneering their exploration and development (Gold, 1985).

A somewhat similar approach was reflected by the position of a senior research officer in a major European corporation in justifying his opposition to the detailed economic evaluation of research proposals and similar innovative activities. His management regards such efforts to reach beyond accumulated experiences as 'similar to an underground stream which wanders unseen for longer or for shorter distances to emerge eventually at some unexpected spot. We do not know where it will emerge, and we do not know when it will emerge, but we want to be there when it does!' In extending his argument, he explained that the company would expect to be first every so often. But even when it was not, its research people would be so familiar with the surrounding technological terrain, and with the paths which had not yet been explored, that it could in most cases duplicate the achievements of the successful pioneers, or develop a reasonable equivalent, in comparatively short order. And, in his view, it is this combination of insurance against catastrophic lags in whatever quarter the competition manages to leap ahead, plus the chance that it will itself achieve leadership with reasonable frequency, that constitutes the soundest justification of the research investments undertaken under such a cloud of uncertainties (Gold, 1973, p.136).

Progress towards such technological improvement goals also requires development of well-informed and responsive managers, technical personnel and workmen. Towards that end, it would be helpful to initiate educational efforts followed by specialized training programs to achieve general understanding of the need for major advances, of their implications for future operations, and of the particular stresses likely to be confronted in the course of absorbing various kinds of innovations. At the same time, careful attention must be given to possible needs for changing existing organizational structures so as to facilitate the effective utilization of major innovations.

For example, the introduction of computerized manufacturing systems requires more effective and continuous integration between computer-aided design and computer-aided manufacturing than is likely to be achieved when each represents a separate organizational unit, leaving the critical point of interaction between them at the periphery of each unit's responsibilities (Gold, 1981, p.31). Such organizational adjustments should also provide for detailed periodic monitoring of progress towards technological improvement goals, including evaluation of the causes of any shortcomings as well as consideration of possible shifts in targets in response to changing pressures and opportunities.

4. EVALUATING SPECIFIC PROPOSED INNOVATIONS

4.1 Suggested Basic Analytical Framework

Private firms are not in business to advance technology or maximize productivity but rather to achieve and maintain attractive rates of profitability. Hence, managements need an analytical framework which facilitates tracing the probable effects of technological innovations beyond productivity to resulting impacts on costs and profits. Conversely, such a framework should also enable managements to trace the sources of unfavorable changes in profits to any underlying alterations in costs and in operating productivity.

Figure 1 presents part of a framework which has been found

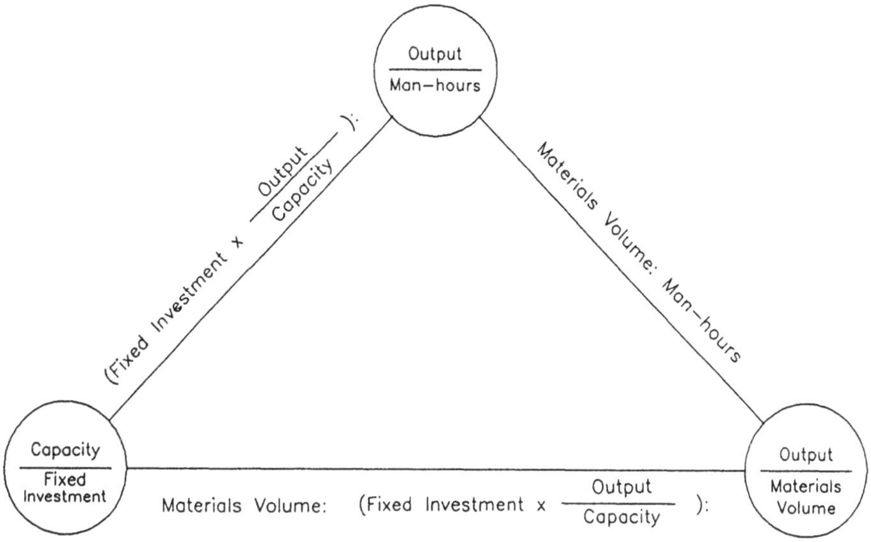

Figure 1. The Network of Productivity Relationships Among Direct Input Factors

useful for these purposes in a wide array of industries. It covers the "network of productivity relationships". This emphasizes that attention must be given not only to the volume of each input required per unit of their joint product, but also to changes in the proportions in which the various inputs are combined. (Note: Fixed investment is compared with the productive capacity which it provides rather than with actual output because the latter is affected by variations in the degree of idleness due to fluctuations in sales.) Specifically, increases in output per man-hour are commonly attributable to utilizing machinery to replace part of the productive contributions formerly made by labor - or to buying more highly processed materials and components which similarly reduce labor's role in remaining processing or fabrication operations. In either case, changes in any of the 6 links in the network require tracing repercussions through the entire system to ensure its effective re-integration. For example, mechanizing some manual operations would tend to increase output per man-hour and to reduce capacity relative to fixed investment in addition to possibly altering unit material requirements as a result of reduced scrap and reject rates - thereby changing factor proportions as well.

In order to explore the prospective effects of such changes in productivity relationships on unit production costs, attention must be turned to accompanying changes in the prices of these input factors, as shown in Figure 2 (see also Gold, 1982, Chapter 3). For example, a 10% increase in output per man-hour, if accompanied by an increase in hourly wage rates of only 5% (an unusually favorable response according to extensive studies), would reduce unit wage costs by 5%. But if such wages account for 20% of costs, total unit costs would tend to decline by only 1%. Even this sharply diminished benefit may be unavailable, however, for one must now ask how the change in output per man-hour was achieved. If it involved the purchase of higher-priced materials or the introduction of more capital goods, their effects on their respective unit costs would have to be weighted by their respective cost proportions to determine resulting changes in total unit costs. Thus, the cost effects of technological innovations often deviate from expectations based only on anticipated changes in physical productivity relationships.

But neither is reducing total unit costs the primary objective of private firms. Changes in the rate of profits on total investment, the one fundamental criterion of performance, are the product of changes in five **interacting** factors: product prices less total unit costs determine average profits per unit of output; and the other factors include the proportion of total investment allocated to fixed investment, the productive capacity provided by that fixed investment, and the rate of capacity utilization, as shown in the equation below:

$$\frac{\text{Profit}}{\text{Total Inv't}} = \left(\frac{\text{Product Value}}{\text{Output}} - \frac{\text{Total Cost}}{\text{Output}}\right) \times \left(\frac{\text{Output}}{\text{Capacity}}\right)\left(\frac{\text{Capacity}}{\text{Fixed}}\right)\left(\frac{\text{Fixed Inv't}}{\text{Total Inv't}}\right)$$

Average Product Prices	Total Units Costs	Capacity Utilization	Productivity of fixed Investment	Internal Allocation of Inv't

Figure 2. Productivity Network, Cost Structure, and Managerial Control Ratios

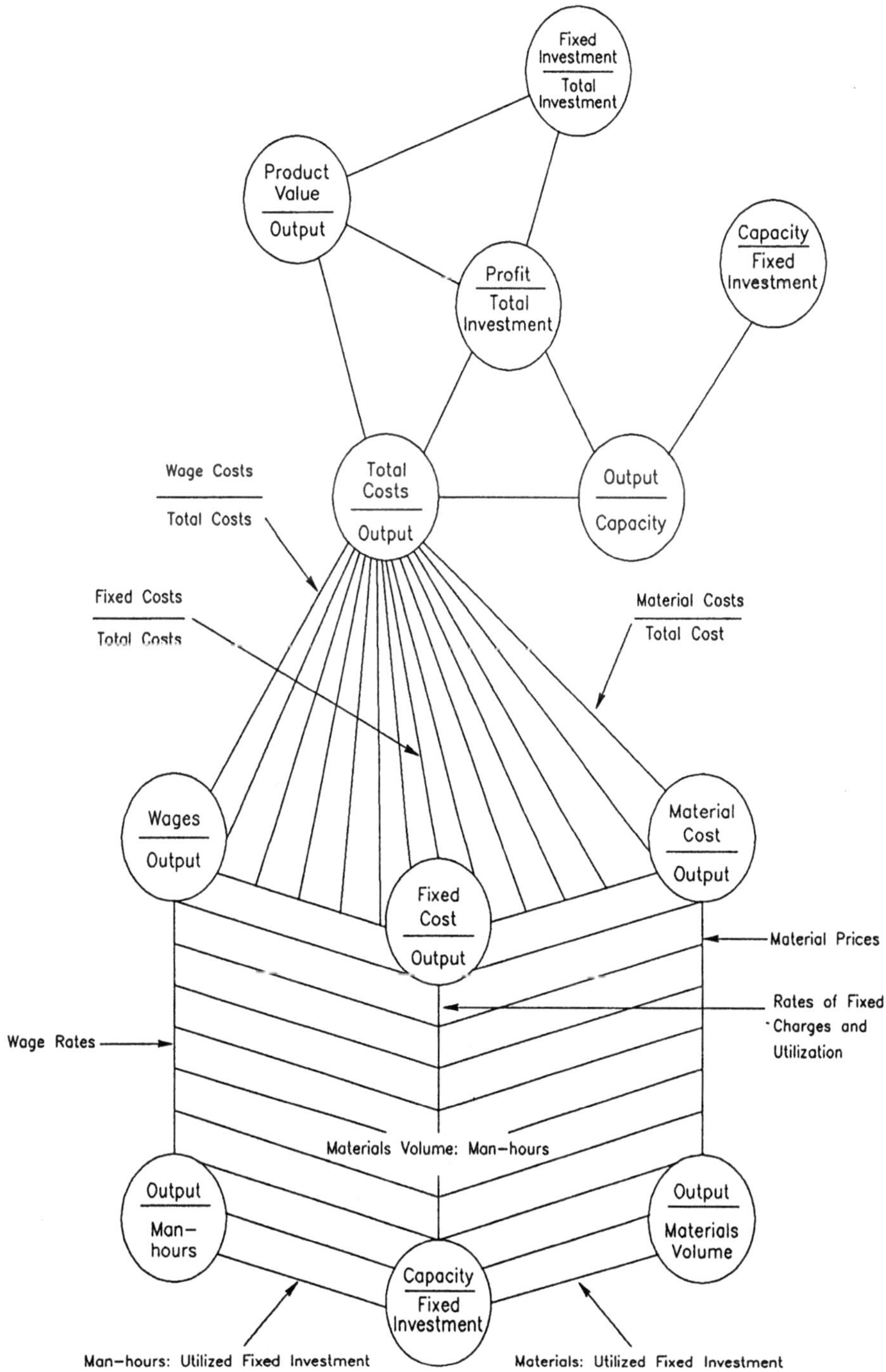

It must be recognized, therefore, that technological advances may also improve product capabilities, which may permit the firm to increase prices or to expand market share in addition to affecting total unit costs, capacity utilization and the productive capabilities of its capital facilities (Gold, 1982). This underlines the point made earlier concerning management planning in section 3.2.

Our extensive research on the development and utilization of major technological innovations also offers three additional suggestions to management. To begin with, maintaining technological competitiveness requires a continuing program of upgrading rather than some single surge effort. It must also be recognized that learning to extract the fullest potential contributions of major technological innovations to increase plant effectiveness often takes several years longer. Hence, as will be discussed later, capital budgeting evaluations of such investment proposals must reach far beyond the currently widespread criterion of "net present value".

4.2 Some Experiential Guides to Improving Pre-adoption Estimates of Performance

Efforts to develop sounder evaluation processes - which may increase adoption rates for some innovations while reducing them for others - require fuller awareness of the criteria which are commonly employed, how they are estimated, and how they are weighted in formulating final evaluations. Only by identifying such elements, and then appraising the vulnerability of the means employed to help guide decisions, can attention be drawn to the relative weaknesses of technological compared with economic estimates of prospective results - as a basis for seeking to improve both.

For example, our field research suggests that the most common sources of errors in estimates of the expected technological benefits of innovations center around:

(1) under-estimating the time needed to achieve effective functioning of the innovations, often by a considerable

margin;

(2) over-estimating the average utilization rate as a basis for appraising benefits;

(3) under-estimating the need to make adaptive adjustments in the preceding and subsequent operations of an integrated production operation - or in the re-allocation of orders and support resources between a new facility and older facilities devoted to similar operations; and

(4) under-estimating the problems and costs of gaining labor acceptance of associated changes in tasks.

Economic evaluations of the expected effects of technological innovations are frequently based on a wide array of simplifying assumptions, among which the following seem to be the most common and most influential:

(1) that expected reductions in man-hours per unit of output will be accompanied by roughly comparable reductions in unit wage costs;

(2) that expected reductions in material requirements per unit of output will yield parallel reductions in their unit costs;

(3) that resulting cost savings can be carried over into increased profits.

But the first two assumptions ignore the tendency for changes in unit input requirements to **interact** with factor prices. Thus, increases in output per man-hour often engender comparable increases in wages per man-hour, whether through piece rates or through trade union demands, thus tending to minimize expected reductions in unit wage costs. Moreover, trade unions often resist the lay-offs made possible by technological innovations. Indeed, our research reveals that some companies in the U.S. no longer permit the inclusion of expected wage cost savings in justifications for proposed capital projects, on the grounds that these all too frequently prove unrealizable.

In the case of reductions in material requirements per unit of output, the effects on their unit costs seem to be more variable. When such reduced inputs are attributable to tighter specifications of quality or dimensions, price increases may offset the expected cost savings. When such reduced inputs have been made possible by an innovation which is being adopted by competitors as well, the expected cost savings may be accentuated as a result of the depressing effect on the price of such materials due to widespread reductions in demand.

But cost savings attributable to shifts to lower-priced materials may shrink in time as competitors also turn to such substitutes and thus increase the demand for them along with attendant prices. Estimates of expected savings in unit material and unit wage costs are also often erroneous because of the tendency to base them on current input prices instead of taking account of recent trends in such prices. Thus, even demonstrable reductions in the quantity of unchanged materials per unit of output may serve only to slow the rate of continuing increases in such unit material costs instead of yielding actual reductions (Gold, 1975).

It is also very important to recognize that the increasing diffusion of cost-saving innovations under competitive conditions tends to generate reductions in product prices as producers struggle to maintain market shares. Thus, the profit margins of early adopters are likely to undergo progressive shrinkage over time. In short, each of the foregoing considerations stresses the importance of seeking to estimate the changing pattern of favorable and unfavorable economic effects period by period over the expected life of the innovation instead of simply multiplying current estimates of annual savings by the expected economic life of the undertaking.

In turn, these perspectives necessitate facing up to the serious inadequacies of the long range economic forecasts which are at the core of capital budgeting evaluations of innovational and other major project proposals. Attendant margins of error are likely to be quite large even in forecasts for entire industries, to say nothing of the even greater hazards involved in forecasts for individual firms and even single plants (Gold, 1974). It would seem useful to consider minimizing dependence on statistical and econometric forecasting in favor of intensive analysis of the specific pressures and opportunities which lie ahead; and to recognize that success is more likely to result from alert recognition of, and effective adjustment to, the inevitable emergence of unexpected developments than on the correctness of the forecasts based on past trends.

4.3 Some Needed Modifications in Capital Budgeting Assessments

Prevailing approaches to evaluating proposals to adopt technological innovations commonly involve estimating their prospective contributions to cost savings and revenue increases as compared with current operations and then comparing the resulting gain in profits with the additional investment required. But such an essentially static perception leads to reliance on the wrong criteria. If significant changes from the present seem likely in any of these primary determinants of future competitiveness, decisions about prospective innovations might be quite different from those counseled by comparisons only with current performance.

In addition, it is important to recognize the serious negative bias involved in evaluating major technological innovations on the basis of the net present value of expected future returns. If such returns are discounted at 15-20% annually, major projects which take even three years to build, "debug", and bring to high levels of utilization would invite rejection in comparison with recent high, virtually riskless money market yields. But such a criterion is dangerously myopic because of its failure to consider the effects on long term competitiveness and profitability of resulting rejections of successive technological advances (Gold, 1979). Hence, it would seem necessary to shift such capital budgeting evaluations from maximizing net present value to what I have called a "continuing horizons" approach. This involves recognition that most major innovations achieve only gradually increasing utilization and benefits, and experience an extended period of reasonable competitiveness and attractive contributions, only to succumb in time to yielding progressively declining returns.

In order to maximize average returns on a continuing basis, therefore, it is necessary to plan for the periodic introduction of new sources of improved competitiveness, recognizing that they will not yield attractive short term returns, and despite the tendency to underestimate the net present value of their eventual contributions engendered by discounting them at high interest rates. Accordingly, one would choose an array of capital projects such that, when the overlapping time paths of expected net returns from each are aggregated, and the "net present value" is calculated as of **successive** three year intervals, the results promise to safeguard acceptable longer term as well as short term rates of profit (Gold, 1975, p.24).

5. APPRAISING INNOVATIONAL EFFECTS

There is limited published research evaluating the actual effects of technological innovations after they have been installed, in contrast to the considerable literature on the initial adoption decision. However, post-installation evaluations offer several potentially important contributions to the management of innovational processes. To begin with, such appraisals could provide a direct comparison of results with the expectations which led to adoption decisions. Even more important, comprehensive evaluations could explore the specific loci and causes of deviations between expectations and results, thus indicating the relative accuracy of various component estimates and also identifying any variables which were ignored. Moreover, such evaluative efforts may reveal insensitivities in the performance measurement system to innovational impacts, thus counseling changes in order to minimize consistent anti-innovation biases.

5.1 Some Shortcomings of Current Evaluation Efforts

Only a limited array of post-installation evaluation methods have been uncovered as a by-product of our field research. Hence, the following judgments must be regarded as preliminary and tentative indications of possible shortcomings.

Most of the 'make good', 'follow up' and 'post-audit' evaluations examined concentrated primarily on simply measuring actual results, including: the costs of acquisition, construction and installation relative to budget; the acceptability of technical performance; and resulting operating costs. Except for comparisons with allowed budgets and expected total unit costs, few methods were characterized by comprehensive comparisons of actual input requirements, factor prices, product quality and price, output levels and other aspects of performance with the respective estimates which led to the decision. Especially glaring is the common failure to consider the time required to achieve the 'acceptable' levels of performance before evaluation efforts tend to be initiated relative to expectations. Also disappointing is the virtual absence of any systematic efforts to 'learn from results' as a means of improving ex ante estimates of the effects of prospective technological innovations in the future.

Our explorations suggest that formal evaluation efforts, except for comparisons of actual expenditures with allowed budgets, seem to be much less common in respect to very large projects, especially when their results seem unfavorable. In explaining such lapses, the two most common reasons given were: that each such project is necessarily unique and hence evaluations would have no feedback value in considering future projects; and that there was no interest in 'spilled milk' or in 'beating dead horses'.

One of the common limitations of post-installation evaluative efforts has been reliance on an overly restricted framework of considerations. Thus, a major part of the decision-making process involves choosing among a variety of available technological and nontechnological means of meeting the most urgent needs of the firm at the time of decision. But a comparison of actual results with expectations in respect to the final choice made throws no light on the accuracy of the evaluations which led to the discarding of the other alternatives considered. Nor do such comparisons reveal the effects of having adopted innovations which succeeded in easing what were considered urgent needs at the time of decision at the cost of neglecting other needs which proved more serious over time. In short, there has been an almost complete failure to evaluate the substructure of evaluations that determined which of the alternative means of dealing with specified needs should be adopted.

Our studies also suggest that another cause of vulnerable

findings has been the apparent pressure for biased evaluations. In the case of very large projects, the tendency to seek out and to emphasize favorable aspects of results seems to be attributable to a concern that negative judgments would reflect on high level officials and be resented by them. Such biases are also often built into the evaluation process because allocation of such responsibilities to the officials deemed to have the relevant expertise involves reliance on those who also participated in project proposals and decisions. Thus, technical evaluations are usually left to engineers and various cost estimates to the respective specialists - partly because of the absence of effective internal alternatives and partly to protect the confidentiality of findings. Moreover, those assigned to making evaluations are often led to mute critical judgments lest these inhibit future cooperative relationships with the officials responsible for the project. Indeed, we have not yet encountered any cases of wholly independent evaluations involving technological and economic competence. The seriousness of this problem is indicated by the fact that a senior officer of one of the major steel companies told us that they have abandoned such post-installation evaluations because they were invariably found to be so biased as to render them of dubious value.

A third set of shortcomings of post-installation evaluations arises from the time foci and criteria employed. Early evaluations tend to be inadequate because it is only after the innovation has achieved effective functioning and reasonably high levels of utilization that efforts to maximize realization of its potentials lead to adaptive adjustments in preceding and subsequent operations, and even to possible modifications in product designs and product-mix. Hence, more effective appraisals would require successive evaluations every 6 months for at least 3 years (or even longer if effective functioning has not yet been achieved - as in some cases of continuous casting, for example) to ensure effective determination of practically sustainable performance levels and to ensure coverage of the wider repercussions of the innovation.

In addition, the actual effects of technological innovations are often measured inadequately because cost accounting categories are not revised to reflect important aspects of the innovation's contributions. These may include changes in the quality of the inputs used, in the nature of the processing or fabrication performed, or in the service capabilities of products. Other significant effects which are commonly disregarded include changes in the flexibility of operations which can be performed and changes in the precision with which processes can be controlled. However, these factors are now being considered by some management accountants in their evaluations of advanced manufacturing technology (for examples see Howell et al., 1987; Kaplan, 1984).

Moreover, concentration on the comparison of results with expectations tends to result in inadequate probing of the specific causes of shortfalls. As a result, observed deficiencies are all too readily ascribed to unpredictable or external factors. Such minimizing of internal shortcomings obviously prevents identification of needed targets for improvement efforts.

Finally, attention should be directed to the seemingly universal avoidance of estimates of the innovation's incremental contribution to profitability. This implied recognition of the difficulties in attempting such evaluations, even on the basis of actual **ex post** data, raises even more serious doubts about the usefulness of **ex ante** estimates of such profitability effects as a basis for adoption decisions.

5.2 Some Additional Problems of Evaluation

Efforts to improve evaluation of the post-installation effects of technological innovations are confronted by a variety of questions concerned with how to reduce biases engendered by common production, costing and other practices. For example, means need to be considered for reducing the favorable biases associated with at least five common production practices encountered in our research. One of these involves shifting the most advantageous orders and the best operators to the new facilities, along with granting them top priority in access to ancillary facilities and to repair and maintenance services. Another such practice involves maximizing the utilization rate of the new facilities at the expense of the older facilities. Such biases also result from motivating greater labor efforts and care through providing improved pay incentives and working conditions. Still another source of differential advantage frequently involves improving the quality of work inputs from preceding operations and increasing the standardization of the tasks to be performed by the new facilities. And the question must also be faced of how long additional improvements to the initial installation are to be attributed to the original innovation instead of to subsequent innovations.

An important group of problems relating to proper evaluation of the cost effects of technological innovations concerns whether the following should be treated as increases in the investment embodied in the innovation or as current additions to operating costs:

(a) additional outlays in order to improve the effectiveness with which the new facility functions;

(b) the cost of interruptions to production caused by introduction of the innovation;

(c) the cost of delays before achieving effective functioning of the innovation, including the costs of modifications, 'debugging', training operators and trial runs; and

(d) the costs and outlays involved in readjusting preceding and subsequent operations in order to achieve effective integration with the capabilities of the innovation.

Perhaps the most difficult problems of all involve seeking to disentangle the effects of the innovation from those of a wide array of concomitant developments. Among these, internal developments might include the introduction of other technological and non-technological innovations, as well as changes in management policies relating to prices, marketing, labor relations and other factors affecting competitive position. External developments might include changes in the availability and prices of input factors, technological and other innovations by competitors, and modifications in government regulations affecting the industry.

Finally, because many technological innovations require investments which are likely to be embodied in them for 10-20 years or longer, some attention must be given to the problem of longer term evaluations. Because relevant costs, revenues and net investment tend to change from year to year, evaluations of an innovation's effects would also yield changing results over time, quite possibly involving substantial changes in their favorableness (Skeddle, 1980). Does this mean that all project evaluations should be repeated annually? For how long can their effects be differentiated from the combined impacts of all other developments? Three other questions seem to be even more fundamental:

(a) What would be the significance for current decision-making of learning that some past decisions yielded favorable results in the short-run, but unfavorable results after 5 years - whereas other yielded the reverse pattern of results?

(b) What margins of error are likely to be associated with the 10-20 year estimates of output, costs prices, interest rates and profits used in capital budgeting models - and should estimates subject to wide margins of error be used as the basis for decisions to adopt or to reject technological innovations?

(c) And if preceding questions are answered in the negative, what alternatives are available to management?

5.3 Strengthening Ex Post Evaluations to Help Improve Future Decisions

Past shortcomings of post-installation evaluations have prevented realization of their potential to improve future evaluations of advanced manufacturing technology. Needed revisions include comparisons of the actual results with estimates of: the probable results of having rejected the innovation, or of having delayed its adoption by 1, 2 or 3 years; the expected results at the time of the original decision; and the apparent results of the technological and other innovational decisions made by competitors at the time when this firm made its original adoption decision. Revised approaches should then seek to explore the causes of the differences revealed by preceding comparisons. In particular, it would be instructive to identify which differences were attributable to technological, economic or market factors; which of these represented internal versus external developments; and, finally, which might reasonably have been predicted at the time of the decision and which were clearly unpredictable.

As a result of such more comprehensive insights into the complex patterns and multiple determinants of the actual effects of technological innovations, consideration might be given to enriching the past objectives, coverage and methods of ex ante evaluations. For example, in defining the objectives on the basis of which choices are to be made among alternative innovations, the need might well be recognized to dig beneath the generalized objective of improving profitability and to concentrate more sharply on the specific product, process, cost and other adjustment targets involved in bettering past performance - thereby providing more precise criteria both for choosing among the options being considered and also for evaluating post-installation results. Moreover, instead of merely comparing the relative net benefits of alternative innovations, efforts should be made to clarify the technological and economic assumptions underlying them, and also to indicate the margins of error likely to be involved - including any relevant references to the results of **ex post** evaluations.

Another by-product of attempts to determine the **ex post** effects of technological innovations more effectively may be recognition of the need to change some of the categories commonly used to assess the productivity and cost effects of prospective innovations. Specifically, measures must be designed to take account of changes in input qualities, the nature of processing requirements, the shifting of processing tasks to other operating units, improvements in process flexibility and alterations in product quality. These tend to alter both the physical magnitudes and economic value of productive contributions and yet have been largely or wholly ignored by prevailing measures, which focus solely on changes in input and output quantities, assuming no significant changes in their qualitative attributes.

6. CONCLUDING OBSERVATIONS

1. There has been a dangerously prolonged under-emphasis in a wide array of industries on supporting vigorous development of their technological capabilities. Resulting

serious lags behind continuously advancing technological frontiers have been a major source of declining competitiveness.

2. Overcoming such lags is likely to prove far more difficult than is widely recognized. The major reason is that such recovery efforts require not only heavy investments in new facilities and equipment, but also substantial changes in managerial priorities and policies, in staff capabilities and in organizational arrangements. An additional reason is that during the extended period needed to overcome built-in resistances, to effectuate required adjustments and to build momentum towards catching up, leading competitors may be expected to keep pushing further ahead.
3. Effective efforts to regain technological competitiveness require:
   (a) top management commitment to this objective;
   (b) development of an array of improvement targets based on objective, competent and comprehensive evaluations of the current and prospective technological advantages and disadvantages of each sector of production as well as of each major component of the firm's product-mix.
   (c) establishing organizational arrangements:
      (i) to continue monitoring relative competitiveness and its determinants;
      (ii) to evaluate alternative technological innovations proposed as means of progressing towards established targets;
      (iii) to appraise the performance of adopted innovations periodically as the basis for uncovering developing shortcomings or additional potentials; and
      (iv) to ensure effective integration of various technological improvement efforts so as to maximize mutual reinforcement and to minimize maladjustments.
4. In order to encourage needed management commitment to, and support for, such recognizably long term programs, whose major benefits may not be realized for some years, it would seem necessary to alter the incentives and rewards which in recent years have motivated an emphasis on maximizing

short term profitability (Gold, 1982).

7. REFERENCES

Gold, B. (1973). "What is the Place of Research and Technological Innovations in Business Planning?" **Research Policy**, Summer.

Gold, B. (1974). "From Backcasting Towards Forecasting", **OMEGA: The International Journal of Management Science, 2** (2), 209-223.

Gold, B. (1975). **Explorations in Managerial Economics: Productivity, Costs, Technology and Growth.** London: Macmillan and New York: Basic Books. Japanese translation: Chikura Shobo, Tokyo, 1977, 193-197.

Gold, B. (1979). "Factors Stimulating Technological Progress in Japanese Industries: The Case of Computerization in Steel", **Quarterly Review of Economics and Business**, Winter (Dec.). Spanish translation in **Clenca Y Desarrolo** (Consejo Nacional de Cienca y Technologia, Mexico D.F.), Nov-Dec 1979.

Gold, B. (1981). **Improving Managerial Evaluations of Computer-Aided Manufacturing.** Washington, D.C.: National Academy of Science Press.

Gold, B. (1982). **Productivity, Technology and Capital: Economic Analysis, Managerial Strategies and Governmental Policies.** Lexington, MA: D.C. Heath-Lexington Books.

Gold, B. (1983). "Technological and Other Determinants of the International Competitiveness of U.S. Industries", **Transactions in Engineering Management of the Institute of Electrical and Electronic Engineers**, May.

Gold, B. (1985). "Strengthening the Foundations of Investment Strategy and Capital Budgeting" in **Handbook of Capital Budgeting.** Kaufman, M. (ed.). New York: Dow-Irwin.

Howell, Robert A., James D. Brown, Steven R. Soucy, and Allen H. Seed. (1987). **Management Accounting in the New Manufacturing Environment: Current Cost Management Practice in Automated (Advanced) Manufacturing Environments.** Montvale, N.J.: National Association of Accountants.

Kaplan, Robert S. (1984). "Yesterday's Accounting Undermines Production," **Harvard Business Review**, July-August, 95-101.

Skeddle, R. W. (1980). "Expected and Emerging Actual Results of a Major Technological Innovation: Float Glass", **OMEGA: The International Journal of Management Science**, October, 553-567.

AN ORIENTING THEORY FOR IMPLEMENTING ADVANCED MANUFACTURING TECHNOLOGIES

John Guarino and David Wilemon
Innovation Management Program
School of Management
Syracuse University
Syracuse, New York 13244

ABSTRACT

Technological innovation is arguably the key competitive challenge of the 1990s. Our paper suggests that the selection and evaluation processes for advanced manufacturing technology should anticipate potential implementation problems as part of the justification process. To this end our paper proposes an orienting theory for implementing technological innovation. The key practical issue is whether people are relatively powerless (preserving hierarchy) or empowered (diminishing the importance and function of hierarchy). We trace this key issue through three perspectives. The first, based on leadership skills, preserves hierarchical arrangements. The second, a human relations approach, can preserve or diminish hierarchy depending on whether people are taken as powerless or empowered. It also suggests an approach based on psychological transactions. This exchange approach makes implementation success depend on building and maintaining adequate coalitions among interested parties.

For scholars, our paper addresses a current dilemma in innovation theory-building: should models be stepwise or simultaneous? We propose that innovation researchers consider going from phase models to an action-oriented focus on what is being exchanged between talented, empowered adults. Managers have been making judgments about who is and who should be involved in innovation implementation, what are their goals and preferences, and what levels of power and power differentials are present while researchers appear not to have noticed. We need to understand the implementation process in terms of exchanges among parties who choose and pursue their own purposes.

1. INTRODUCTION

> The organizational hills are full of managers who believe that an innovation's technical superiority and strategic importance will generate acceptance
> (Leonard-Barton and Kraus, 1985).

... and the high technology hills are full of superior manufacturing technologies that failed to produce the expected benefits. Our thesis is that the selection and evaluation processes for advanced manufacturing technology (AMT) should incorporate more than technological and economic analyses. Managers need to anticipate (cf. Meyers, 1989) potential implementation problems as part of the justification for - or rejection of - advanced manufacturing technology.

To this end our paper proposes an orienting theory (Whyte, 1984) to answer Leonard-Barton and Kraus (1985) and Voss' (1985; 1988) call for research on implementing new technologies. According to Whyte (1984), the purpose of an orienting theory is to guide inquiry into an emerging field. In this approach, the investigator follows opportunities to learn even to the point of departing from the original research design. Such inquiries can force revisions of an emerging theory. Our orienting theory, then, is a pragmatic attempt to gain new knowledge about AMT implementation. Theory guides data collection, which in turn helps guide theoretical development. Opportunities for learning take precedence over following a pre-planned research design.

Our unit of analysis is the decision to adopt advanced manufacturing technology by a business firm or business unit in a multibusiness firm. This is one type of innovation-decision (Downs and Mohr, 1976). The innovation itself, AMT, consists of a "bundle of inventions" (Schumpeter, 1934) often developed from the microprocessor (Dean, 1987). Also, AMT is often called "computer integrated manufacturing" (CIM; Dean, 1987).

1.1 Relation to the Innovation Literature

Much of our argument about advanced manufacturing technology parallels the new product development literature. The key difference is that new product development generally does not make implementation a discrete area of inquiry. A new product such as a detergent is produced, purchased by individual users (or rejected), and then diffused into the population of users (or fails to diffuse). A technological innovation, by contrast, often requires buyers to spend considerable time and effort getting it ready for use. Often the users of technological innovations need to work closely with suppliers to achieve successful implementation. Implementation intervenes between the purchase decision and full use.

1.2 Roadmap to the Paper

Our paper first describes the strategic rationale for adopting advanced manufacturing technology. Section two categorizes reasons why managers may be reluctant to or choose not to invest in AMT, suggesting that managers can enact limitations on their own ability to compete (Smircich and Stubbart, 1985).

Section three gives three approaches to implementing advanced manufacturing technology. The key issue is whether people are viewed as relatively powerless (preserving hierarchy) or empowered (diminishing the importance and function of hierarchy). We trace this key issue through three perspectives based on, respectively, leadership skills, human relations concepts, and exchange theory. In the concluding section we recommend a pragmatic, learning orientation to the complex world of AMT implementation.

2. POTENTIAL BENEFITS FROM ADVANCED MANUFACTURING TECHNOLOGY

Dean (1987) finds that AMT benefits start with increased flexibility. According to Dean

> (The) combination of shortened product life cycles and market fragmentation places a substantial premium on

> flexibility in manufacturing. In order to compete in many industries today, and even more in the future, firms will need the ability to produce a wide variety of customized products simultaneously and to abandon production of current products in favor of new ones quickly. Globalization has truly rewritten the rules of competition: in order to be viable competitors, firms must be able to manufacture a rapidly changing mix of high-quality, customized products at very low costs (1987).

Other authors often cite a second benefit: increased productivity (Davis and associates, 1986). Valery (1986) reports that United States industrial productivity has grown at a rate of 5% annually since the recession of the early 1980s. This compares favorably to growth rates of 0.1 % from 1973-1981 and 2% from 1960-1973. It seems reasonable to infer that part of the increase in growth is due to new technology. Ettlie (1988) cites a survey of 700 industrial engineers in which 81% cited new automated equipment as the explanation for productivity improvement (Material handling engineering, January, 1984). One-third of the plant modernization cases studies by Ettlie reported productivity as the most important rationale (1988).

Other benefits reported are enhanced control, accuracy, and customer responsiveness (Jelinek and Goldhar, 1986). These benefits enhance the ability to serve markets as they become increasingly segmented (and ultimately fragmented; Dean, 1987). Under such conditions, variety can drive production efficiency as does high production volume. That is, economies of scope are starting to be of comparable importance to the familiar economies of scale. Successful implementation of new technology creates the ability to serve more, and smaller demand segments than is possible otherwise. Sustainable competitive advantage based on economies of scope (Goldhar and Jelinek, 1983) often depends on microprocessor technology. Calls for revisions in production economics to reflect these new strategic realities are now common (Gold, 1982; Kaplan, 1984; Skinner, 1984).

The literature also suggests that some managers have been slow to make the link between advanced manufacturing technology and sustainable competitive advantage (Buffa, 1984; Hayes and

Wheelwright, 1984; Skinner, 1985). According to Dean

> Perhaps the most consistent theme that has run through the discussion on the decline in American competitive ability, however, is our failure to remain technologically competitive. American firms have historically been slow to implement state-of-the-art technologies, even though many of them have been developed here. This has often cost them dearly, as firms from other countries have used these technologies. In so doing, they have created a wide technological gap, which in turn resulted in disparities in cost, quality, and flexibility. The most recent example of this phenomenon is the reluctance of American firms to implement advanced manufacturing technology, which may be our last hope for global manufacturing viability in many industries (1987).

Ettlie (1985) reports that few managers keep up with this literature. To the extent that this is true, managers may miss the link between new technology and sustainable competitive advantage. We further suggest that even if managers make this link, they may have good reasons to pass up AMT. Informed managers know that successful implementation of advanced manufacturing technology is certainly not automatic and that AMT's full contribution to sustainable competitive advantage is often difficult to achieve.

In light of the above, Hage asks an excellent question: Why do some managers choose to abandon an established business when technology is available that could allow them to successfully compete and continue in business, even though they have the money to purchase it (Hage, 1986)?

3. POTENTIAL COSTS OF ADVANCED MANUFACTURING TECHNOLOGY

Given the strategic arguments for AMT, Hage's question is an interesting one: Why do managers enact (Weick, 1979) limitations on their own ability to compete rather than try new technology? We suggest there are potential costs as well as the benefits discussed above. We classify these costs in terms of reasons to avoid advanced manufacturing technology. There may be economic reasons, workplace control issues, and reasons related to managerial self-esteem and self-justification.

3.1 Economic Reasons to Enact Limits to Competing

Managers do not always respond to market conditions (Hage, 1987). We argue in this section that economic reasons to pass up new technology can develop from both strategic and distributive considerations.

3.1.1 Strategic Reasons

The first criterion for the AMT selection decision is the business unit's strategic thrust. The new technology might not fit, in which case managers might have reason not to try to generate profit in an industry. For example, if managers propose to harvest a current line of business, or enter into a new industry or segment, then they are less likely to adopt AMT in that business.

Another reason to forego new technology is the difficulty of forecasting earnings and cash flows. Some observers suggest that misused financial analysis (Myers, 1984; Reich, 1983) or biased accounting techniques (Kaplan, 1984) can cause managers to look away from long term investment in new combinations of means of production. Dean finds that the future economic benefits of the new technologies are hard to quantify. Managers commonly cannot project return on investment, internal rate of return, net present value, or payback period. In most firms, there are numerous strategic and financial arguments for not investing in AMT. If no one can provide a clear strategic or financial justification, top management will eliminate costly technology projects (1987).

Dean explores the organizational and interpersonal processes that take over in the face of financial uncertainty. No one to our knowledge has written on organizational and interpersonal reasons to <u>reject</u> advanced manufacturing technology. To these we now turn.

3.1.2 Distribution of Value Added

The second economic reason why a manager might enact limits on the business' ability to compete resides in potential problems with competing claims on the fruits of successful innovation. Unsure of the distribution of returns, managers may decline to adopt new technology even if the projected returns are high and AMT is in line with the business' strategic thrust.

Economic issues are tough. Managers are responsible for business performance, which depends in part on conditions outside managerial control. Also, in allocating profits to claimants, people may not get as big a share as they think equitable. Notions of equity (Adams, 1965) suggest the presence of noneconomic reasons for management to decide against advanced manufacturing technology. The next two sections give noneconomic reasons to enact limits on the business unit's ability to compete. We suggest that each develops from attempts to preserve and justify hierarchy. Also, the remainder of our barriers to new technology occur at the plant level.

3.2 Control Reasons to Enact Limits on Competing

Social criteria strongly influence adoption and implementation of new technology. At the level of a national economy, Emery (1982) makes social acceptance of technology more important than its availability in determining Kondratiev Long Waves. Hage (1986; 1987) makes a similar case at the economic sector level. Other authors make this argument at the plant level (Edwards, 1979; Howard, 1985; Noble, 1986). Finally, Walton (1982) suggests that meeting broad human needs such as autonomy (eg, Hackman and Oldham, 1975), social connectedness, meaningful work, and effective voice (Hirschman, 1970) should be, and will be, necessary for applying advanced information technology to white collar work.

This sampling of the literature yields three propositions from which a common theme emerges. The propositions are:

1. People <u>choose</u> what they will do,

2. People <u>choose</u> what technological changes are possible, and

3. People <u>choose</u> what technologies they will not support.

The common theme is that human factors may be as important as a "technological imperative" (Woodward, 1965) in determining who will control work behavior.

All this is problematic to managers who expect their organizations to function hierarchically. Taylor (1911), Goodrich (1920), Rumelt's (1974) deflation of Galbraith's (1967) "technological imperative", and current writing on the NUMMI plant (Holusha, 1989) have in common the assumption that system purposes should come before individual goals. Most managers and management scholars see the issue of workplace control as important, but resolved; by taking a job, a person agrees to take on system purposes.

Managers and scholars may forget that the "components" of a system are people. Managers and management scholars are trained to explain current organizational arrangements (Scott, 1974) using a systems paradigm, which is explicitly hierarchical (Kast and Rosenzweig, 1972). Note that Burns and Stalker's (1961) distinction between organic and mechanistic systems does not matter; both are hierarchical and assume superordinate system purposes. The potential danger here is in assuming that people give up their purposes in favor of system purposes. To clarify this, see the comparison of control in the systems world (labelled "cybernetic") vs. control in a world of empowered adults ("political") in Table 1.

The left hand column of Table 1 gives facets of cybernetic control (Hofstede, 1978). Cybernetic control ultimately depends on a hierarchical approach; higher-level elements decide what lower-level elements will do in pursuing a system-level goal or set of goals. Higher-level people induce lower-level people to comply by offering monetary or social rewards. Power resides at the top of the system (Kast and Rosenzweig, 1972) although it is delegated to the relatively powerless lower levels as the top people see fit. According to Hofstede (1978), cybernetic control systems work

Table 1. Cybernetic Control vs. Political Control

	Cybernetic systems	Political
Source of control	Hierarchical control	Bargaining, negotiated order
Standards	A standard exists	Objectives are missing, shifting, unclear
Performance assessment	Accomplishment is measurable	Accomplishment is not measurable
Feedback	Situations recur; feedback is usable	Events do not recycle; value of feedback is low

Source: Hofstede (1978)

reasonably well when three conditions hold: a performance standard exists, a metric exists to measure performance, and situations tend to recur so feedback supplies useful information. In sum, cybernetic control works in a relatively simple world.

In contrast, political control (see right hand column of Table 1) is based on the idea of a "negotiated order" (Strauss, 1978) in which working relationships result from bargaining processes. Note that goal consensus (agreement on ends) is not needed. For action to occur, it is enough to agree on means (Lindblom, 1959; 1980). In the political perspective people choose their own sets of goals and bargain to reach them. The analyst might not know either what empowered people are trying to accomplish or whether they have accomplished it. Finally, political control is appropriate when events do not recur; feedback from previous experience is not so valuable as in a simpler world.

Clearly the prerequisites for a standard, measurable accomplishment, and usable feedback cannot be met except in a very simple world. The world in which managers must make advanced manufacturing technology decisions is complex and is trending toward increased complexity. For example, Leonard-Barton (1988) reports that new technology changes performance criteria and control arrangements at all organizational levels. Hage (1987) and Child (1987) write that the most important issue in implementing microelectronic technology may be whether the goal of innovation is to control and deskill people. In the same vein, Zuboff (1988) suggests that it is possible to informate (place knowledge in people) as well as automate (place knowledge in machines). If knowledge yields power, then the decision between informating and automating is a power issue. Hage (1987) and Child et al. (1987) also consider whether knowledge will be invested in people or machines. Manz (1983) suggests that to succeed in a complex world, managers need to empower their subordinates; our paper suggests that it may be time for a more extreme solution.

In implementing advanced manufacturing technology, managers need to consider how they "get in the way". According to Chairman James Houghton of Corning, Inc., "Today's better-educated workers want to have the power to control their own workplace lives. This should be encouraged. In our factories and businesses we have hundreds of teams that spot trouble and fix it at the source without supervisors and top managers, like me, interfering" (Houghton, 1989).

One way to "stay out of the way" is to consider that new high technology combinations of productive resources have high information content. General or functional managers who lack advanced training in mechanical or electrical engineering, physics, chemistry, etc. may have little understanding of the technologies used in producing their products (Hetzner et al., 1986). To the extent that this condition is real, questions of workplace control are relevant, open, and negotiable. To try to retain control over workplace behavior by keeping users of technology separate from the necessary information reduces efficiency and courts failure.

Howard (1985) reports an evocative case although it does not arise in a manufacturing firm. Eastern Airlines' management, desiring to keep workers away from the information needed to control their jobs, put the on-off switch used to gain access to a computer's memory under lock and key. One worker, a qualified locksmith, provided a key. Both sides continued this petty gamesmanship through several dreary rounds. Eighteen months after the flap started, the workers presented the issue to Eastern's President Frank Borman, who wrote a memo. Our point is that no matter who "wins" this kind of battle, the corporation, its owners, its managers, its supervisors, its workers, and its customers incurred the cost.

3.3 Self-Esteem and Limits on Competing

Managerial self-esteem may be the key issue in deciding on advanced manufacturing technology. Rapid technological change is a permanent feature of our lives (Kimberly, 1987). We argue that managers are biased to fear change. In a rapidly changing world, depending on complex technologies having high information content and facing demanding subordinates, previously successful managers often face unwanted "opportunities" that can bring about loss of current rewards and jeopardize future rewards. From such fear would develop a natural concern for any increase in the complexity of managerial life. Differently put, managerial fear of change can be a major barrier to implementing advanced manufacturing technology. While some managers may wish for a simpler world with pliable employees, this is not the world of AMT.

3.3.1 Fear of Complexity

The environment is complex and managers often lack information and insight into what is happening. Such complexity can arrest logical thought. If managers believe the world of new technology will be complex, then it is difficult to justify the effort to implement it. Managers may not see potential aids in AMT implementation. Wanting not to be "shown up" or symbolically "killed off" (eg, Slater, 1966), managers may evaluate new technology negatively, decide to stick with current technology and thereby enact limits on their business unit's ability to compete. The theme is: the world of new technology is complex and complexity is often threatening.

Anticipating that implementing advanced manufacturing technology is difficult, and not knowing whether it will produce success, managers may let competitors innovate even to the point of taking away customers. Addressing the difficulty of implementing AMT, Peters says:

> (I)mplementation of the new integrated information technology-based systems is much more difficult than anyone dreamed. For one thing, it turns out that the installation of such systems is not primarily a matter of technology. It is a matter of organization. <u>Every power relationship, inside and outside the firm, is affected by the installation of the new information technology systems</u>. The failure of so many elaborate new systems (in many arenas the failure rate is estimated at well over 90 percent) typically results from a failure to think through the bare-knuckle issues of power redistribution. In a word, to implement such systems, hierarchy must be destroyed (1987).

The dilemma is: managerial self-esteem hinders the needed deemphasis on hierarchical arrangements, but hierarchical thinking doesn't work very well in implementing new technology. This logic underlies Morgan's recommendation to avoid hierarchical forms in the workplace (1988).

Peters' second theme is that managers should empower people. The sticky point is that employees who understand microelectronic technology and its use are already empowered. In such cases we suggest management might do well to get out of their way.

4. APPROACHES TO IMPLEMENTING MICROELECTRONIC TECHNOLOGY

In this section we propose three approaches to advanced manufacturing technology implementation. The first is a leadership skills-based approach which may preserve and work within the constraints of hierarchy. The second, construction of a more humane work system, can preserve or diminish hierarchy. The final perspective, an exchange approach, is inherently nonhierarchical.

4.1 Leadership Skills

The first approach to implementing advanced manufacturing technology is to use special leadership skills to preserve current role configurations by appealing to, for example, superior technological expertise or charisma. Under the norm of rewarding individual merit, superior leadership skills support claims to a higher organizational position. Authors mining this vein advise managers to empower subordinates (Kanter, 1983; Peters, 1987), "get out of their way" (Peters, 1987), or develop skills in "remote management" - that is, shaping subordinates' behavior by creating "appropriate" organizational values and cultures, avoiding direct operational control, and getting actively involved only when subordinates cannot solve a problem (Morgan, 1988). The gist of these prescriptions is that "higherups" should cede control as needed to subordinates. This also assumes that managers know when to cede control and when to reclaim it.

4.1.1 Expertise

Skinner emphasizes that managers must have a mental picture of the physical processes in the production system if they are to manage it intelligently. We join him in believing that expertise is central to implementing advanced manufacturing technology. Since AMT implies new combinations of factors of production (Schumpeter, 1934), implementing it commonly requires expertise which few may possess. Almost everybody may be "lost" so the managerial claim to "expert" authority may be problematic.

4.1.2 Charisma

The second approach that preserves hierarchical arrangements is to use charisma (attractiveness, personality, or French and Raven's "referent power" (1959)) to elicit from the lower levels of the organization a high degree of commitment to AMT implementation (Walton, 1985). This is based on the assumption that the lower levels are inferior in some way (note that higherups are called "superiors") and can be manipulated for the collective good.

In line with the theme of manipulating lower level members of the organization are three recommendations for managers: foster ambiguity (Quinn, 1979), manage meanings for subordinates (Smircich and Morgan, 1982), and take on special roles (Maidique, 1980). While most managers and management scholars see ambiguity as dysfunctional, Urwick gives a classic statement of its positive aspects for managers: "It is much easier to secure acceptance of a scheme if no one understands what is involved" (1942). First, Quinn (1980) suggests that a manager might use ambiguity to get subordinates to commit to courses of action. This enhances the likelihood that each alternative will have an advocate so good alternatives will be less likely to die (Quinn, 1980). Second, ambiguity from above avoids win-lose conflicts for subordinates: if top management doesn't identify with a particular alternative, then nobody has to be on the wrong side. Finally, ambiguity can be a lubricant that allows successful, talented, powerful people to work together over longer periods of time.

Selznick (1957) and Smircich and Morgan (1982) make the case for managing the meaning of reality for subordinates. These authors make leadership the selection of meanings from an essentially ambiguous, perhaps meaningless, stream of experience. Those meanings that are selected and retained form a collectively constructed interpretation of reality which enables members to communicate in a common language and thus organize their interrelated behaviors. Under this shared interpretation of reality, organizational activities may proceed.

Selznick's "institutional leader" creates vision and values for subordinates. If needed, this leader concocts a "socially integrating myth" to preserve the system. Smircich and Morgan (1982) make a similar case. Managers should frame issues for subordinates, select interpretations of reality, create visions, etc. Managing meanings invokes the skills and attitudes toward subordinates of the priest or shaman. A goal here is to preserve hierarchical power and control.

The final approach to developing charisma is to take on roles such as innovation champion, sponsor, orchestrator (Maidique, 1980), resource person, or model (Molinari, 1988). Thinking in terms of roles can be fruitful. For example, role thinking surfaces possible "resistor roles". It also surfaces the question of top management involvement; if advanced manufacturing technology needs someone to run interference, then a high degree of senior management involvement is indicated. If management is likely to intrude inappropriately, then a low degree of involvement would be indicated.

Thinking in terms of roles may also have a downside. If we take the notion of a self-organizing technology development team (eg, a "skunkworks" team) seriously, then role thinking may unnecessarily constrict members' choices to perform the tasks involved in innovating. Thinking and acting in terms of roles can also block creativity. If the prescribed expectations for a person in a role are too specific, the person may feel pressure to "work to rule" with dysfunctional consequences for the overall team effort.

The leadership skills approach is likely to succeed. People often deskill themselves by withdrawing, by feeling helpless, and by acting passively, thereby getting a superior to take on a leadership role (Gemmill, 1986). The result is that they often do act like powerless "sheep who have lost their way" (Rioch, 1959). The leadership skills approach also maintains and reinforces hierarchical arrangements. If it seems unlikely to succeed, then the manager might try to implement advanced manufacturing technology by creating a more humane workplace. For example, human relations management sets out to meet workers' social needs and thereby elicit enhanced performance (Miles, 1975). We posit that the key issue is whether managers treat subordinates as powerless sheep (hierarchy survives) or as empowered adults (the dysfunctional consequences of hierarchy diminish).

4.2 Make a Humane System

Two theoretical umbrellas comprise this perspective: human factors and human relations. Human factors attempts to match job facets, which may change when advanced manufacturing technology is adopted, with facets of people. Majchrzak (1988) and Ettlie (1988; Chapter 7) describe the impact of technological change on jobs. The approach is explicitly hierarchical; people fit into roles in an organizational system. These roles, however, can and often do change with the introduction of new technology.

In contrast, the human relations school is concerned with people in groups. Management's task is to create a good climate. Supposedly satisfaction will enhance, or at least correlate with, performance in implementing advanced manufacturing technology. Of the several aspects of human relations theory, we suggest that transactional analysis (Harris, 1967) can be particularly useful in implementing new technology. Transactional analysis uses the notion of ego states to organize thinking about exchanges between people. If a person is in the "child" state, another can form and maintain a hierarchical relationship by operating in either an "adult" or "parent" state. A dependent, powerless child makes a "good" subordinate.

In contrast, an empowered adult makes a poor subordinate in the usual sense. The very notion of hierarchy can interfere with getting adults to implement new technology. Nor is the notion of "empowering" people very helpful; those who understand microelectronic technology and its implementation can empower themselves. The question for the future of hierarchical arrangements, then, is whether people are empowered or powerless. It is easy to "manage" the powerless by issuing commands. The more difficult problem is to comprehend and deal with empowered adults who understand microelectronic technology.

4.3 Exchange

One answer may lie in negotiating to settle the terms under which interlocked behaviors (Weick, 1979) required to implement advanced manufacturing technology will occur. We view these interlocked behaviors as transactions or exchanges of behavior (Emerson, 1981).

> The secret of working with talented people is not to impose your will on them (Bob Newhart, in Brady, 1989).
>
> The secret of working with talented people is you cannot impose your will on them (the Unknown Exchange theorist).

A secret of implementing new technology may be that, although a business unit needs people to debug the equipment (Jonsson, 1987), one cannot order talented people to succeed. We consider technology implementation success as a function of negotiating outcomes. We will show how an exchange perspective can address several challenges in implementing AMT (Leonard-Barton and Kraus, 1985).

4.3.1 A World of Exchange

The key notion of the exchange perspective is that talented, empowered adults do not act like elements in a production function. Life among adults is transactional. It is often impossible to specify what will constitute compliance with the terms of an agreement, so a typical agreement states the limits on allowable behaviors. The key notion is that each party's expectations concerning the other's future behavior should match the other's perceptions of those expectations and vice versa.

4.3.2 Application to Innovation Implementation

Leonard-Barton (1988) concludes that it is useful to cast the process of innovation implementation in terms of negotiating toward mutual benefit for all involved. If life is inherently transactional, then one should think in terms of bargaining to form a coalition to implement the innovation (Barnard, 1938; March and Simon, 1958). Peters (1987) prescribes adding value through people at all levels; the exchange perspective suggests they should share in the added value. Peters quotes Roger Smith on innovation: "(F)orty guys can't do something that three people can do. It's just a law of human nature" (Peters, 1987). While one may empathize with Mr. Smith's frustration, this is obviously no "law of human nature". If it were, then GM could innovate by taking 37 of the forty away from the innovative project. Thinking about people in terms of "laws of human nature" obscures understanding.

> Learning fails because events are caused and consequences felt by different organizations (Pressman and Wildavsky, 1973).

Prospective benefits to the system or to top management do not persuade talented, empowered people to implement technological change. Prospective benefits to themselves are more likely to persuade them to help. If today's winners face a chance to become tomorrow's losers, they are not likely to favor innovation. This suggests a cost-benefit analysis for all people who will be involved in implementation. In Mr. Smith's case, if the forty don't expect net benefits from the innovation, they might not help implement it. They may have good reasons for nonperformance. To get them to perform might require providing rewards for innovation. To say that forty cannot do what three can do is silly; to say that they <u>won't</u> do what three can do may be the first step toward improved management of advanced manufacturing technology implementation. People do what yields valued rewards.

> The manager (who resisted) never voiced his opinion, since there was little rational basis for it, but his resistance effectively stalled the project. (Leonard-Barton and Kraus, 1985).

4.3.3 Resistance to Change

Leonard-Barton and Kraus (1985) suggest that resistance to change is a function of the prospective costs and benefits to the potential resistor. In a classic article, Paul Lawrence (1954) hypothesizes that people resist not technological, but social change. The sociotechnical systems perspective implies that people resist both social and technological change (Trist, 1981). To overcome resistance to change, one may do well to concentrate on social and technological changes simultaneously (Ettlie, 1988; Leonard-Barton, 1988; Peters, 1987; Taylor et al., 1986). An exchange perspective suggests how to do so; evaluate the cost/benefit balance for all participants. Find out what people want as inducements to participate, give it to them, and get out of their way. What managers experience as resistance to change may be rational adult functioning. We suggest that managers anticipate it in the selection and evaluation stages to enhance the prospects for implementing new technology.

4.3.4 Other Challenges

Leonard-Barton and Kraus (1985) note resistance to change as the first challenge in innovation implementation. Second, the presence of multiple internal markets suggests an approach based on negotiating a series of exchanges. Third, deciding how intensely to internally promote innovation can be analyzed in terms of expectations. Overpromotion of advanced manufacturing technology can violate people's expectations and perhaps boomerang. Finally, managers desiring to implement advanced manufacturing technology must deal with hedgers or fence-sitters. Leonard-Barton and Kraus (1985) recommend managing performance criteria to persuade hedgers to help implement innovation. In sum, Leonard-Barton and Kraus seem to have in mind a bargaining approach to changing existing arrangements.

4.3.5 Conclusions

Peters (1987) and Leonard-Barton and Kraus (1985) made major contributions to justifying and implementing advanced manufacturing technology. Many of Peters' propositions related to leadership and empowering people fit in an exchange worldview. While he appears to get caught in the "systems thinking" trap his ideas have considerable merit. Hierarchical organizations can indeed sabotage progress, so some diminution of hierarchy may be needed. Peters exaggerates the point by concluding that the destruction of hierarchy is necessary. An exchange-focused analysis suggests, however, that he errs in concluding that managers need to empower people and that the course of innovation implementation is unpredictable (1987). In the world of advanced manufacturing technology people can empower themselves; the manager does not need to. Workplace democracy may be only a microprocessor away.

The exchange perspective also offers levers to influence the course of AMT implementation and deal with several issues raised be Leonard-Barton and Kraus (1985). The exchange approach shows how to surface and manage resistance, points out the need for congruency between the parties' expectations and mutual perceptions, and offers a window into making exchanges regular by forming contractual relations. Our paper offers this as a useful adjunct to the technological and economic analyses involved in new technology decisions.

5. CONCLUSIONS AND IMPLICATIONS

Our main conclusion is that there is no fully satisfactory theory of innovation implementation. The primary implication for managers is to prepare for a learning experience. Advanced manufacturing technology implementation is a hard field for the researcher or the manager to enter because the theory is neither fully developed nor clear. Little deductive theory exists (for notable exceptions see Rogers, 1983; Voss, 1985, 1988). On the other hand, many induced sets of prescriptions are available (see Leonard-Barton and Kraus, 1985; Peters, 1987; Ettlie, 1988; Morgan, 1988). It is unclear how generalizable the induced theories are. Nonetheless, prescriptions and orienting theories can help managers who need to map the complex world of microelectronic technology.

Our paper offers an orienting theory encompassing three perspectives on implementation. The first, based on leadership skills, preserves hierarchical arrangements. The second, a human relations approach, can preserve or diminish hierarchy depending on whether people are viewed as powerless or empowered. It also suggests an approach based on psychological transactions, more fully developed as the exchange worldview. The exchange approach, our third approach, offers a difficult, but realistic and intelligible, avenue for learning to enhance the likelihood of implementation success. Treating organizational life as a series of exchanges develops informed judgment by surfacing what costs and benefits are going to whom, thus decreasing complexity. The exchange perspective makes implementation success depend on building and maintaining adequate coalitions among interested parties. If a coalition cannot be formed, or if it disintegrates, then success is unlikely.

5.1 Implications for Managers

The complex world of microelectronic technology is here. Our orienting theory deals with AMT in ways no current theory does. Our theory offers three major reasons why managers fail to purchase advanced manufacturing technology even when the economics are favorable. Managers decide not to purchase new technology for strategic and distributive reasons, for workplace control reasons, and due to the real and persistent threats the new technology may pose to managers. Our theory then suggests that managers approach implementation in three generic ways (using leadership skills, human relations techniques, or negotiating tactics). We suggest that treating implementation as a process of ongoing negotiation is the most approach most likely to lead to successful implementation.

We offer one additional suggestion for managers; consider the possibility that past learning can be an impediment to success. If experience is the best teacher, perhaps one's next experience will be an even better teacher than past experience. Initial AMT implementation can be an "information rich" exemplar for future efforts even if there are failures along the way.

We have offered an orienting theory - a pragmatic attempt to discover what works and what doesn't in situations in which managers must act. This effort will be a success if it is found to surface the important issues involved in implementing advanced manufacturing technology during the selection and evaluation stages. It is no substitute for the best judgment of the participants. As Urwick puts it, "If in their practical application (these principles are found) repugnant to reason and common sense, then let reason and common sense prevail" (1942).

5.2 Implications for Future Research

For the scholar, our paper addresses a current dilemma in innovation theory-building: should models be stepwise or synchronous (Ettlie, 1988)? Our theory complements the phase models which abound in innovation research by a non-stepwise, pragmatic approach which focuses in on the critical actors, their goals and preferences, and the important power issues that can arise in AMT implementation. To date, the scholarly literature has not converged on a single stepwise model of innovation implementation and seems unlikely to do so (Van de Ven and Rogers, 1988). While the advent of inexpensive computing power enables us to construct an almost infinite variety of contingency theories by augmenting stepwise models with recycles, feedback loops, and new linkages (cf. Venkatraman, 1989), we should begin to confront the possibility that we may find many different implementation processes which relate to each other in unpredictable ways. This suggests that our theoretical ideals cannot be met; an action-oriented, diagnostic approach to innovation implementation may be the best available. Given the complexity of the stepwise approaches, our non-stepwise, pragmatic approach may comprise an idea whose time has come. In this view, an interesting question is whether combining the leadership skills, human relations, and exchange approaches to planned change would work better than any one approach.

This is not to call for a moratorium on further development of phase models. It is to propose that innovation researchers consider going from phase models to an action-oriented focus on what is being exchanged between talented, empowered adults. Implementing AMT is arguably the key competitive challenge of the 1990s. Researchers (like managers) need to make judgments about who is and who should be involved in innovation implementation, what are their goals and preferences, and what levels of power and power differentials are present. Managers have been doing this while researchers appear not to have noticed. We need to get close to power differentials and to understand the action in terms of bargains (explicit and implicit) among the involved parties who choose and pursue their own purposes. There is evidence that they do this in ways that detached scholarly analysis may not be able to tap. The world of advanced manufacturing technology implementation is up close and personal. Our paper proposes a pragmatic orienting theory for use in finding out what works in a world that is progressively less safe for armchair analysts.

6. REFERENCES

Adams, J. S. (1965). "Inequity in social exchange," **Advances In Social Psychology, 2**. New York: Academic Press.

Barnard, C. I. (1938). **The Functions Of The Executive.** Cambridge, MA: Harvard University Press.

Brady, J. (1989). "Interview With Bob Newhart," **Parade**, April 9, 19.

Buffa, E. S. (1984). **Meeting The Competitive Challenge.** Homewood, IL: Dow Jones-Irwin.

Burns, T. and G. M. Stalker. (1961). **The Management Of Innovation.** London: Tavistock.

Child, J. (1987). "Managerial Strategies, New Technology, and The Labor Process," **New Technology As Organizational Innovation.** Cambridge, MA: Ballinger, 141-177.

Child, J., H. D. Ganter, and A. Kieser. (1987). "Technological Innovation And Organizational Conservatism," **New Technology As Organizational Innovation**, Cambridge, MA: Ballinger, 87-116.

Davis, D. D., and Associates (1986). **Managing Technological Innovation.** San Francisco: Jossey-Bass.

Dean, J. (1987). **Deciding To Innovate.** Cambridge, MA: Ballinger.

Downs, G. W., Jr., and L. B. Mohr. (1976). "Conceptual Issues In The Study Of Innovation," **Administrative Science Quarterly, 21**, 700-714.

Edwards, R. (1979). **Contested Terrain.** NY: Basic Books.

Emerson, R. M. (1981). "Social Exchange Theory," **Social Psychology.** NY: Basic Books, 30-65.

Emery, F. E. (1982). "New Perspectives On The World Of Work," **Human Relations, 35**, 1095-1122.

Ettlie, J. E. (1985). "The Implementation Of Programmable Manufacturing Innovations," **Dissemination And Implementation Of Advanced Manufacturing Processes.** Washington, D. C.: National Science Foundation.

Ettlie, J. E. (1988). **Taking Charge Of Manufacturing.** San Francisco: Jossey-Bass.

French, W. L. and B. Raven. (1959). "The Bases Of Social Power," **Studies In Social Power.** Ann Arbor, MI: Institute for Social Research of the University of Michigan, 150-167.

Galbraith, J. K. (1967). **The New Industrial State.** Boston: Houghton-Mifflin.

Gemmill, G. (1986). "The Mythology Of The Leader Role In Small Groups," **Small Group Behavior, 17** (1), 41-50.

Gold, B. (1982). "CAM Sets New Tules For Production," **Harvard Business Review, 60** (6), 88-94.

Goldhar, J. D. and M. Jelinek. (1983). "Plan For Economies Of Scope," **Harvard Business Review, 61** (6), 141-148.

Goodrich, C. L. (1920). **The Frontier Of Control.** NY: Harcourt, Brace, and Rowe.

Hackman, J. R. and G. Oldham. (1975). "Development Of The Job Diagnostic Survey," **Journal Of Applied Psychology, 60**, 159-170.

Hage, J. (1986). "Responding To Technological And Competitive Changes: Organization And Industry Factors," **Managing Technological Innovation.** San Francisco, CA: Jossey-Bass, 44-71.

Hage, J. (1987). "Reflections On New Technology And Organizational Change," **New Technology As Organizational Innovation.** Cambridge, MA: Ballinger, 261-276.

Harris, T. A. (1967). **I'm OK - You're OK.** NY: Harper and Row.

Hayes, R. H. and S. C. Wheelwright. (1984). **Restoring Our Competitive Edge: Competing Through Manufacturing.** NY: Wiley.

Hetzner, W. A., J. D. Eveland, and L. G. Tornatzky (1986). "Fostering Innovation: Economic, Technical, and Organizational Issues," **Managing Technological Innovation.** San Francisco: Jossey-Bass, 239-255.

Hirschman, A. O. (1970). **Exit, Voice, and Loyalty.** Cambridge, MA: Harvard University Press.

Hofstede, G. (1978). "The Poverty Of Management Control Philosophy," **Academy Of Management Review, 3**, 450-461.

Holusha, J. (1989). "No Utopia, But To Workers It's A Job," **New York Times**, January 29, 1-10.

Houghton, J. R. (1989). "The Age Of Hierarchy Is Over," **New York Times**. September 24.

Howard, R. (1985). **Brave New Workplace**. NY: Viking Press.

Jelinek, M. and J. D. Goldhar. (1986). "Maximizing Strategic Opportunities In Implementing Advanced Manufacturing Systems," **Managing Technological Innovation**. San Francisco: Jossey-Bass, 220-238.

Jonsson, S. (1987). "Limits Of Information Technology For Facilitating Organizational Learning," **New Technology As Organizational Innovation**. Cambridge, MA: Ballinger, 217-234.

Kanter, R. M., 1983. **The Change Masters**. NY: Simon and Schuster.

Kaplan, R. S. (1984). "Yesterday's Accounting Undermines Production," **Harvard Business Review**, 95-101.

Kast, F. E. and J. E. Rosenzweig. (1972). "General Systems Theory: Applications For Organization and Management," **Academy Of Management Journal, 15**, 47-465.

Kimberly, J. R. (1987). "Organizational and Contextual Influences On The Diffusion Of Technological Innovations," **New Technology As Organizational Innovation**. Cambridge, MA: Ballinger, 237-259.

Lawrence, P. R. (1954). "How To Deal With Resistance To Change," **Harvard Business Review, 32** (3), 49-57.

Leonard-Barton, D. (1988). "Implementation As Mutual Adaptation Of Technology And Organization," **Research Policy, 17**, 251-267.

Leonard-Barton, D. and W. A. Kraus. (1985). "Implementing New Technology," **Harvard Business Review, 63** (6), 102-110.

Lindblom, C. E., 1959. "The Science Of Muddling Through," **Public Administration Review, 19**, 79-88.

Lindblom, C. E., 1980. **The Policy-making Process**. Englewood Cliffs, N.J.: Prentice-Hall.

Maidique, M. A. (1980). "Entrepreneurs, Champions, and Technological Innovation," **Sloan Management Review, 21** (2), 59-76.

Majchrzak, A. (1988). **The Human Side Of Factory Automation**. San Francisco: Jossey-Bass.

Manz, C. C. (1983). **The Art Of Self-leadership**. Englewood Cliffs, NJ: Prentice-Hall.

March, J. G. and Simon, H. A. (1958). **Organizations.** NY: Wiley.
Materials Handling Engineering. (1984). "Manpower and Automation: Keys to Productivity," January, 21.

Meyers, P. W. (forthcoming). "Nonlinear Learning In Large Technological Firms: Period Four Implies Chaos," **Research Policy.**

Miles, R. E. (1975). **Theories Of Management.** NY: McGraw-Hill.

Molinari, J. M. (1988). "Perceptions Of The Degree Of Implementation Of Microcomputers and Associated Factors: A Study Of Public Radio Stations," Unpublished Dissertation, Graduate School of Management, Syracuse University.

Morgan, G. (1988). **Riding The Waves Of Change.** San Francisco, CA: Jossey-Bass.

Myers, S. C. (1984). "Financial Theory and Financial Strategy," **Interfaces, 14** (1), 126-137.

Noble, D. F. (1986). **Forces Of Production.** NY: Alfred A. Knopf.

Peters, T. J. (1987). **Thriving On Chaos.** NY: Harper and Row.

Pressman J. L. and A. B. Wildavsky. (1973). **Implementation.** Berkeley, CA: University of California Press.

Quinn, J. B. (1979). "Technological Innovation, Entrepreneurship, and Strategy," **Sloan Management Review, 20** (3), 19-30.

Quinn, J. B., 1980, **Strategies For Change.** Homewood, Ill.: Richard D. Irwin.

Reich, R. B. (1983). **The Next American Frontier.** NY: Times Books.

Rioch, M. J. (1959). "All We Like Sheep... (Isaiah 53:6): Followers and Leaders," **Psychiatry, 14**, 258-273.

Rogers, E. M. (1983). **Diffusion Of Innovations.** NY: Free Press.

Rumelt, R. P. (1974). **Strategy, Structure, and Economic Performance.** Cambridge, MA: Harvard Business School.

Schumpeter, J. A. (1934). **The Theory Of Economic Development.** Cambridge, Massachusetts: Harvard University Press.

Scott, W. G. (1974). "Organization Theory: A Reassessment," **Academy Of Management Journal, 17**, 242-254.

Selznick, P. (1957). **Leadership In Administration.** N.Y.: Harper & Row.

Skinner, W. (1984). "Operations Technology: Blind Spot in Strategic Management," **Interfaces, 14** (1), 116-125.

Skinner, W. (1985). **Manufacturing: The Formidable Competitive**

Weapon. NY: Wiley.

Slater, P. E. (1966). **Microcosm**. New York: John Wiley.

Smircich, L. and G. Morgan. (1982). "Leadership: The Management Of Meaning," **Journal Of Applied Behavioral Science, 18**, 257-273.

Smircich, L. and C. Stubbart. (1985). "Strategic Management In An Enacted World," **Academy of Management Review, 10**, 724-736.

Strauss, A. (1978). **Negotiations**. San Francisco, CA: Jossey-Bass.

Taylor, F. W. (1911). **The Principles Of Scientific Management**. New York: Harper and Brothers.

Taylor, J. C., P. W. Gustavson, and W. S. Rogers. (1986). "Integrating The Social and Technical Systems Of Organizations," **Managing Technological Innovation**. San Francisco: Jossey-Bass, 154-186.

Trist, E. L. (1981). "The Sociotechnical Systems Perspective," **Perspectives On Organization Theory and Behavior**. New York: Wiley, 409-418.

Urwick, L. F. (1942). **The Elements Of Administration**. NY: Harper and Brothers.

Valery, N. (1986). "High Technology: Clash Of The Titans," **The Economist, 300**, 7460.

Van de Ven, A. H., and E. M. Rogers. (1988). "Innovations and Organizations," **Communication Research, 15**, 632-651.

Venkatraman, N. (1989). "The Concept Of Fit In Strategy Research: Toward Verbal and Statistical Correspondence," **Academy Of Management Review, 14**, 423-444.

Voss, C. A. (1985). "The Need For A Field Of Study Of Implementation Of Innovations," **Journal Of Product Innovation Management, 4**, 266-271.

Voss, C. A. (1988). "Implementation: A Key Issue In Manufacturing Technology: The Need For A Field Of Study," **Research policy, 17** (2), 55-63.

Walton, R. E. (1982). "New Perspectives On The World Of Work: Social Choice In The Development Of Advanced Information Technology," **Human Relations, 35**, 1073-1084.

Walton, R. E. (1985). "Toward A Strategy Of Eliciting Employee Commitment Based On Policies Of Mutuality," **HRM: Trends and challenges**. Boston: Harvard Business School Press, 35-68.

Weick, K. E. (1979). **The Social Psychology Of Organizing**. Reading, MA: Addison-Wesley.

Whyte, W. F. (1984). **Learning From The Field**. Beverly Hills:

Sage.

Woodward, J. (1965). **Industrial Organizations: Theory and Practice.** London: Oxford University Press.

Zuboff, S. (1988). **In The Age Of The Smart Machine.** NY: Basic Books.

INDUSTRIAL ADOPTION AND USE OF PROGRAMMABLE CONTROLLERS

Professor Richard C. Dorf
Johann Sandholzer
Department of Electrical Engineering and Computer Science
Graduate School of Management
University of California, Davis, CA 95616

ABSTRACT

Programmable controllers, originally used as a replacement for relay logic, have evolved into an important tool in factory automation and Computer Integrated Manufacturing. Programmable controllers are used mainly for production and process control. The benefits, obstacles and the success factors associated with their usage were examined through a user survey. Most respondents used additional advanced technology such as Computer Aided Design or Automated Quality Control in conjunction with programmable controllers. For the majority of the respondents, programmable controllers were integrated within manufacturing automation. Approximately half of the respondents did not undertake a feasibility study to evaluate programmable controllers which, surprisingly, did not influence the time needed for implementation or contribute to the difficulties encountered.

The greatest benefits derived were higher reliability, less downtime and increased flexibility for product changes. The difficulties most commonly encountered were lack of in-house expertise, high cost of equipment, problems with installation and integration and lack of technical support through vendors/distributors/manufacturers. The evaluation criteria normally used were payback period, repair time, downtime and productivity. The most desirable features of programmable controllers are presented, and the paper concludes with a series of policy recommendations and implications.

1. INTRODUCTION

This study examines the usage and adoption of a multipurpose automation tool, the programmable controller. Usage of programmable controllers is concentrated on the factory floor, the lowest level of the control hierarchy in Computer Integrated Manufacturing (CIM) (Sugiyama and Williams, 1988). The programmable controller is a highly developed, user-oriented product with the software and hardware option for communications interfacing with supervisory control systems (Bentley, 1987).

Hierarchical control systems address the need for efficiency, productivity and quality in the automated factory. A top-down design based upon hierarchical, distributed control architecture can be represented by a pyramid model (Struger, 1985). The utility of control systems at each level is summarized as:

plant level: overall planning, execution and control

center level: scheduling production and management information

cell level: coordinating multiple stations

station level: controlling real-time devices

machinery and process level: interfacing to real-time devices

The programmable controller is an essential part in the lower levels of the hierarchical and distributed control architecture. Over the last two decades, it has evolved from simple relay replacement to a stand-alone programmable controller, and is now an integrated part of an overall manufacturing or process control system (Smith, 1988). Since distributed control systems tend to be expensive, they are only used for large processes. The programmable controller can be a good choice for small batch control projects and batch processes (Norris and Kross, 1988; Jacic, 1988). Programmable controllers are no more than part of a total production system, and it is the impact on the efficiency of the system as a whole which is important.

Before examining the adoption and usage patterns of programmable controllers, we should briefly consider what a programmable controller is. The National Electrical Manufacturers Association (NEMA) defines programmable controllers as "digitally operating electronic apparatus which uses programmable memory for internal storage of instructions for implementing specific functions such as logic, sequential timing, counting, and arithmetic control through digital or analog modules, various types of machines or processes" (Raia, 1986 and IEEE Standard Dictionary of Electrical and Electronics Terms, 1977). One example is a stand alone unit controlling the process flow in a food plant. The programmable controller has its program stored in an internal memory and is used to mix different ingredients by opening and closing valves at different times. The input data from the temperature sensors are in analog form and the counter of the weighing scale gives its information in digital form. The system can become more complicated if more programmable controllers are combined and connected within the same control hierarchy (peer-to-peer).

This study focuses on programmable controllers, partly because of their own importance, and partly because their usage is a broader indicator of the extent of adoption of advanced manufacturing technology. The next section provides background information for the study. It is followed by a discussion of programmable controller user characteristics including the plant size, industry and applications. Next details on the initial usage of programmable controllers, including the use of feasibility studies and time required for installation are given. This is followed by sections on the level of usage across the sample. Then the benefits and difficulties expected and experienced are examined. A short case study follows. Finally, the implications of the facts revealed by the study are considered and analyzed in order to determine the actions required for successful adoption of programmable controllers.

2. THE STUDY

The study took the form of a mail survey of existing programmable controller users and one case study interview of a successful user. The survey was undertaken during July and August 1988 and the follow-up interview was conducted in November 1988.

Since no list of all programmable controller users was available, the sample for the users survey was based on a number of different sources: names and companies from articles in automation magazines, recommendations, members of the Instrumentation Society of America and participants of conferences on automation. The survey was sent to 150 persons, and the response rate was 25% - a normal range for surveys of this category. Altogether, the effective sample amounted to 38 responses. The responding users represent a wide variety of industries and plant sizes.

The subjects covered by this survey included the characteristics of programmable controller users, factors that influenced the programmable controller's introduction, the way the first programmable controller was set up and the application for which the controller was employed. In addition, respondents were asked about the benefits and difficulties expected and experienced, the degree of success achieved, the productivity measures used, and plans for the future.

3. CHARACTERISTICS OF PROGRAMMABLE CONTROLLER USERS

3.1 Size

The most striking characteristic of programmable controller users is that the plants at which they are employed tend to be large. More than four-fifths of the respondent users work in plants that employ more than 100 people and one-third work in plants employing more than 1,000 people (Table 1). Only one in ten users has annual company revenues of less than $ 10 million, and more than one-fourth of the responding companies has an annual turnover of more than $500 million.

Table 1. Distribution of Responding Programmable Controllers Users by Employment Size (N = 38)

Range	Percent
1 - 49	7
50 - 99	10
100 - 499	36
500 - 999	10
1,000 - 4,999	28
5,000 +	7

There are several possible reasons for the skew toward large users in the sample. First, larger firms more often have a scale, range and sophistication of production operations suitable for the use of programmable controllers. Second, larger firms typically have the funds to buy programmable controllers. Finally, larger firms tend to be more aware of the opportunities and benefits offered by programmable controllers.

3.2 Industry

The responding programmable controller users represent a wide range of industries (Table 2). Users in the automotive, chemical and aerospace industries tend to be particularly large in terms of plant size and company turnover. However, those in the energy and plastics industries are often smaller.

A large proportion of programmable controller users are companies which themselves manufacture programmable controllers or integrate them into systems. Such companies are probably more likely to be aware of the advantages of programmable controllers and to possess the expertise to use them efficiently. Only a very few of the responding users have programmable controllers in experimental development and in prototype plants.

The programmable controller is rarely used without associated technology. Programmable controllers were most often used in

Table 2. Distribution of Responding Programmable Controller Users by Industry (N = 38)

Industry	Percent
automotive	16
aerospace	8
chemical, petrochemical	16
electrical, electronics	11
energy	13
food, tobacco	13
machinery	13
metal goods	8
plastic products	5

conjunction with CAD, followed by automated testing and quality control (Table 3). The use pattern may affect the evaluation and progress of programmable controllers.

Table 3. Respondents' Use of Other Advanced Technology (N = 38)

Technology	In Use (%)	Planned(%)
CAD	67	13
Computer Aided Manufacturing	33	26
NC Programming Systems	33	7
Robots	28	18
Automated Testing and Quality Control	49	31
Computer Control of Machine Groups	41	31
Automated Storage	18	21

Other responses: Automated Guided Vehicles, Digital Process Control, AI, Fiber Optics, CAM Information System

4. INITIAL USE OF PROGRAMMABLE CONTROLLERS

4.1 Source of Initiative

The source of the initiative to use programmable controllers varies greatly from company to company, reflecting the great diversity in forms of company organization. Often the level of the company where the idea of using programmable controllers originated and the level where the decision to use controllers is executed are not the same (Table 4).

Table 4. Location of Origination and Decision to Adopt Programmable Controllers (N = 35)

Level	Original Idea		Final Decision	
	No.	Percent	No.	Percent
company board of directors	0	0	2	6
business general manager	5	14	6	17
plant management	8	23	12	34
department level	22	63	15	43

Nearly two-thirds of all respondents indicated that the source of the original idea was at the department level. A possible explanation is that programmable controllers are considered less important by upper management and less expensive than other advanced manufacturing technology. Almost one-fourth of the adoption decisions originated at the plant management level, while about one-third of the final decisions were made at the plant management level. In many organizations, the final decision was made at a level higher than that of the source of the original idea. In larger plants, decision making tended to be more broadly distributed.

4.2 Feasibility Studies

About one-half of the responding companies did not evaluate programmable controllers with a feasibility study before their introduction, while about one-third did, using in-house staff. In one out of eight cases, the study was done by suppliers or vendors (Table 5).

Table 5. Organization Undertaking Feasibility Study (N = 38)

	No.	Percent
no study	18	47
study, using the in-house staff	12	32
study, using the supplier or vendor	5	13
study, using an independent consultant	2	5
study, with the help of a group of companies	1	3

The surprisingly high incidence of firms who reached an adoption decision without a feasibility study could be attributed to confidence in the firm's expertise to introduce programmable controllers successfully. However, the lack of a detailed evaluation could also reflect an unawareness of potential problems. In the case of smaller plants, it could reflect more informal decision-making arrangements and the absence of a need to make a formal study as part of the process of securing approval from the head office. Whatever the reasons, employing a feasibility study did not increase the time needed for a plant to place programmable controllers into commercial use (Table 5). The most frequently adopted approach was the use of an in-house feasibility study. In any case, the responses to the introduction of programmable controllers among those who did not perform a feasibility study do not appear to be very different from the responses of those who did carry out feasibility studies.

An in-house study was a particularly common approach in the

larger plants, which more often have the technical expertise to make this practicable. The next most common approach, followed by about one-eighth of all responding users, was to have the supplier of the programmable controller undertake the feasibility study. The obvious disadvantage of this approach is obtaining an appraisal from a biased source. However, this disadvantage was felt to be more than offset by the advantage of drawing on the supplier's relevant expertise and familiarity with the particular equipment envisaged. However, another reason for the surprising result could be that programmable controller users interpret the term "feasibility study" in different ways (e.g., some refer to only the programmable controller, while others refer to the whole system).

4.3 Approach of Installation

For more than four-fifths of all responding programmable controller users, the assembly and installation was planned and executed by corporate engineering staff. This approach costs less initially, but can lead to difficulties if the firm lacks the specialist expertise to get the programmable controller running or to integrate it effectively into the automation concept. On the other hand, no major technical differences exist between the turnkey installations performed by manufacturer/vendor or the consultant. Both approaches are more expensive initially, but the user can benefit from the experience of the supplier or consultant and reduce the risk of things going awry.

4.4 Time Needed for Installation

Only one programmable controller user surveyed did not have its first programmable controller in commercial use. The time required for successful implementation normally was found to be less than four months. Interestingly, the implementation period did not seem to be related to whether a feasibility study had been undertaken (Table 6).

Table 6. Time Period Required for Users to Achieve Successful Implementation (N = 38)

Time Period (months)	No.	Percent
1	4	14
2	8	29
3	10	36
4	2	7
5	0	0
6	3	11
9	1	4
mean = 3 months		

5. NUMBER OF PROGRAMMABLE CONTROLLERS

5.1 Number of Programmable Controllers Per Plant

Nearly one-fifth of the respondents are using one to five programmable controllers. At the other end of the scale, about one-fourth have more than 100 programmable controllers in use (Table 7). To a certain extent, these differences can be explained in terms of the different lengths of time that programmable controllers have been in use by the individual companies.

Table 7. Number of Programmable Controllers Per Plant (N = 38)

Range	Percent
1 - 5	17
6 - 25	34
26 - 100	23
100 +	26

5.2 Industrial Activities Using Programmable Controllers

The top three reported activities for programmable controllers are about equally distributed in the process control, production control and material handling categories. The rest are split among the different activities: chemical processing, energy management, metalworking and food processing (Table 8). This distribution reflects a widespread use of programmable controllers. A more detailed analysis about the distribution of the number of programmable controllers separated in different industries could not be made due to an absence of the necessary data.

Table 8. Industrial Activities Using Programmable Controllers (N = 38)

Industrial Activity	Percentage Use
process control	58
production control	53
material handling	50
energy management	32
metalworking	26
chemical processing	18
food processing	16

5.3 Size of Programmable Controllers

The distribution of the size of programmable controllers in our survey is almost uniform. The future trend is towards mid-size programmable controllers, which consist of medium (129-892 Input/Output) channels and small (65-128 I/O) controllers (Table 9). This distribution suggests that the number of small to medium size programmable controllers will increase in the future, a result which is in disagreement with the report from Frost & Sullivan (Hayner et al., 1988).

Table 9. Size of Programmable Controllers Used (N = 38)

Size (in Input/Output channels)	Adopted Now (%)	Plan Purchase in Future (%)
micro(<64)	55	11
small(65-128)	47	26
medium(128-892)	68	29
large(893+)	42	16

6. APPLICATIONS

6.1 Operations Controlled by Programmable Controllers

Process monitoring/control was indicated as an application of programmable controllers by more than two-thirds of all respondents. The next most frequent application is data acquisition, followed by assembling and machining applications (Table 10).

Table 10. Programmable Controller Operations (N = 38)

Controlled Operations	Percent Using PC
process monitoring/control	71
data acquisition	47
assembling	32
machining	26

Most of the larger plants report using more than one application, which indicates the broad use of programmable controllers. These results are only slightly different from the results obtained by **Design News** in its survey (Bulkeley, 1987).

6.2 Control Scheme

Nearly two-thirds of all responding plants are using the programmable controller as a stand-alone device. Next in importance are continuous and distributed control schemes, each used by about four-tenths of the respondents. Batch processes and hierarchical control were indicated by about one-third of the respondents (Table 11).

Table 11. Control Scheme of Programmable Controllers (N = 38)

Overall Control Scheme	Percent
stand-alone	61
distributed	45
continuous	42
batch	32
hierarchical	32
peer-to-peer	21
centralized	21

While batch process control is used most often in chemical industries, continuous processes are used most often in the automotive industries. Stand-alone systems were rated higher than peer-to-peer systems (i.e., programmable controllers that are combined and connected within the same system), which are more difficult to control.

6.3 Uniform Protocol

Almost half of all responding plants did not have a uniform protocol in order to connect different programmable controllers or other devices that have a defined list of commands. About one-fourth had a uniform protocol, and the rest were planning to establish one. About 42% of the users who had no Local Area Network (LAN) did not have a uniform protocol either. However, 40% of the responding companies were using a LAN, and the rest were planning to install one.

One-fourth had not integrated the programmable controller into their manufacturing automation. The same number reported plans for integration, and the others reported that they had already included it in their conceptual planning.

These results are similar to the findings of a survey made by **Design News**, in which 62% of the programmable controllers of the responding companies were not compatible with General Motors Manufacturing Automation Protocol (MAP) specifications. In this survey, 73% of the responding companies' systems did not incorporate a LAN (Bulkeley, 1987).

7. BENEFITS

7.1 Main Benefits Experienced

In general, the benefits achieved from the use of programmable controllers have been very similar to those expected by the users. The rating of the benefits were very high, so that the small deviation between benefits expected and benefits achieved can only be seen in a very relative manner. The greatest benefits experienced were increased reliability, reduced downtime and an increased flexibility for product changes. Rapid fault diagnosis received the next highest rating (Table 12).

An overall benefit index was developed, which is the sum of the benefit ratings for all responding users for each question. To illustrate the computations, consider the following example. If no answer was given, a zero was entered. These sums represent an Index or Total Score. This same approach is used for Tables 12-15, 21, 24 and in Figure 1.

The overall trend was toward expectations that were greater than actual achievements. However, it is important to note that the expected results were not usually based on a feasibility study.

Rating	Benefits
1	negative
2	none
3	little
4	many
5	many unexpected

Sample question
Increased flexibility for product changes:

Respondent	Expected	Achieved
1	4	5
2	3	3
.	.	.
.	.	.
.	.	.
N	4	3
.	.	.
.	.	.
.	.	.
38	.	.
	Expected Index	Achieved Index

Table 12. Benefits Experienced by Responding Programmable Controller Users (N = 38)

Benefit Description	Total Score Expected	Total Score Achieved
increased flexibility for product changes	118	117
greater reliability, less downtime	119	112
minimizing downtime (faster fault diagnosis)	103	100
lower maintenance costs	97	98
improved quality, more consistent products	89	87
higher throughput	77	74
decreased bottlenecks in process flow	68	69
lower labor costs	63	66
lower material costs, less waste	64	65
maximize plant capacity	67	65
lower energy costs	61	64
less capital tied up in work in progress	61	61

8. DIFFICULTIES

8.1 Difficulties Expected Prior to Adoption

The difficulty most expected prior to adoption was a lack of in-house expertise in working with programmable controllers. The second most common concern was problems with installation and integration in the system, which may relate to the lack of in-house expertise (Table 13). The third expected difficulty was the high cost of equipment associated with the use of programmable controllers within a system.

Table 13. Principal Difficulties Expected by Programmable Controller Users Prior to Adoption of Programmable Controllers (N = 38)

Description of Difficulty	Total Score
lack of technical in-house expertise	84
problems with installation and integration	71
high cost of equipment	70
plant personnel do not accept system	69
lack of technical support	64
insufficient reliability, maintenance	53
inadequate after-sales support from suppliers	53
no advantage over existing equipment	51
no advantage over hard automation	45

8.2 Difficulties Actually Experienced

The difficulties experienced were rated in the following order: high cost of equipment, lack of technical support, problems with installation and integration and lack of technical expertise (Table 14).

The difficulty ranked next was opposition from the shop floor, which fell into the position that was expected (Table 13). Based on additional comments on the questionnaires, it appears that opposition to the new equipment arises in some cases from middle or upper management. Inadequate after-sales support from the supplier was not one of the main concerns. The lowest rated difficulties were: no advantage over existing equipment, no

Table 14. Difficulties Experienced by Responding Programmable Controller Users (N = 38)

	Total Score	Before	After Installation*
high cost of equipment	86	2	1
lack of technical support	72	4	2
problems with installation and integration	69	3	3
lack of technical in-house expertise	67	1	4
plant personnel do not accept system	63	5	5
no advantage over existing equipment	53	8	6
inadequate after-sale support from suppliers	53	6	6
insufficient reliability, maintenance	52	6	8
no advantage over fixed automation	45	9	9

* Ranking of difficulties most expected and most experienced, respectively.

advantage over fixed automation and insufficient reliability and maintenance. Sufficient reliability, as reported by the respondents, was reached after a period of adapting and correcting the programmable controller inside the system. The two other low-rated difficulties can be explained by the relatively long time period programmable controllers have been commercially available.

9. EXPECTED SUCCESS OF ADOPTION

Although the general experience of using programmable controllers has been successful, some have been more successful than others. It is therefore of importance to review the findings of the survey in order to identify which factors contribute to the higher success rate. The following facts should be taken into account concerning these factors. First of all, the factors are not mutually exclusive. Secondly, as with most generalizations, they do not apply in all cases. Finally, our sample size is not large enough to provide for a completely reliable statistical result.

Nevertheless, an examination of the factors associated with greater than average success does provide reasonable indication of the probability of success given different characteristics and courses of action.

9.1 Managerial Measurements

The survey provides six measures that can be used as indicators of success: expected payback period, return on investment, productivity, increased profit, lessened repair time and downtime and lower warranty costs for goods produced by programmable controllers. The last measure should be an indicator for the quality of the products after using the programmable controller. These measures should help to determine the advantages of the use of programmable controllers as well as reflect the usage of different measurements within industries (Table 15).

Table 15. Performance
(N = 38)

Performance Index	Total Score
expected payback period	149
lessened repair time and downtime	149
productivity	139
return on investment	131
increased profit	120
lower warranty costs	98

In the following observations a summation index ESA, synonymous for Expected Success of Adoption, will be used. It is calculated from the sum of the ratings for six performance factors for all plants. The summary performance indicator ESA is calculated the following way:

Ratings Scale	Respondent 1 answered: Performance Factor	Rating
1 not worthwhile	expected payback period	5
2 not worthwhile	return on investment	4
3 marginally worthwhile	productivity (output/hour/system)	3
4 fairly worthwhile	increased profit	3
5 very worthwhile	lessened repair- and down-time	4
	lower warranty costs	2
	ESA	21

The next respondent had a sum of 23, and so on. The average of the responses is recorded in Tables 16, 17, 18, 19 and 20.

9.2 Plant Characteristics

Larger plants (in excess of 1000 employees) were more successful in the adoption of programmable controllers. No meaningful difference existed between the success rates of responding plants with sizes of 1-99 and 100-999 employees (Table 16).

Table 16. Size of Plant vs. ESA

Number of Employees Per Plant	Average of ESA*
1 - 99	23.2
100 - 999	23.3
1000+	26.8

* For all plants in the category.

Some differences were found among industries. For example, the automotive and aerospace industries, which tend to use a wide variety of advanced automation tools, had on the average a higher ESA. Surprisingly, the chemical and petrochemical industries had the lowest ESA despite their tendency to have a large number of employees and high revenues (Table 16, Table 17).

Table 17. Industry vs. ESA

Industry	Average of ESA*
automotive & aerospace	27
metal, plastics,machinery,food	25
electrical, energy	24
chemical, petrochemical	23

* For all plants in the category.

9.3 Introduction of Programmable Controllers

Undertaking a feasibility study had negligible influence on the ESA. Table 18 shows that the ESA for the group not using a feasibility study was higher than the ESA for the group using an in-house feasibility study. This unexpected result could be explained by the fact that those using a feasibility study had higher expectations. It could not be determined whether the undertaking of a feasibility study and the difficulties experienced were interrelated.

Table 18. Feasibility Study vs. ESA

Feasibility Study Status	Average of ESA*
none undertaken	25.4
in-house	23.2
other (vendor, group)	25.8

* For all plants in the category.

The plants that had programmable controllers for the longest period reported deriving the most benefits from them and finding them most worthwhile. Possible reasons include better estimate of benefits and learning curve (Table 18, Table 19).

Table 19. Period Acquired vs. ESA

Time Period	Average of ESA*
before 1979	27.2
1980 - 1983	24
after 1984	22.6

* For all plants in the category.

9.4 Number and Application of Programmable Controllers

Plants with a higher number of programmable controllers on the whole tended to perform better (Table 20). Clearly, the advantage of experience on the learning curve led to improved implementation.

Table 20. Number of Controllers vs. ESA

Frequency	Average of ESA*
1 - 25	21.4
25 - 99	24
100+	26.4

* For all plants in the category.

10. IMPORTANT FEATURES FOR USING PROGRAMMABLE CONTROLLERS

As indicated by the responding users, the most important feature in using programmable controllers was relay-ladder language programmability (the programming used for relays is available for programmable controllers). Almost the same level of importance was attributed to fault tolerance (reliable against influences by electronic fields, etc.) and rapid fault location. The next most desired feature was self diagnosis and self detection of errors. Flexible manufacturing system capability and standardization of interfaces also achieved high ratings (Table 21). Surprisingly, the availability of application software was rated as the least important feature.

Table 21. Important Features for Using Programmable Controllers (N = 38)

Feature Index	Total Score
relay-ladder language programmability*	153
fault tolerance and rapid fault location**	152
self diagnosis and self detection of errors	149
flexible manufacturing system capability	145
standardization of interfaces	142
increased flexibility for product scope	118
increased flexibility for product mix	117
cost per channel	112
memory capacity	109
availability of application software	93

* programming language based on relay logic symbols
** robust against environmental influences (e.g., electromagnetic fields) and fast fault location.

The variability in the level of important factors appears to be dependent upon the size of the plant and the length of time from adoption of a programmable controller (Table 22 and Table 23).

A variety of factors influence the selection of specific programmable controllers (see Figure 1). The two most important reasons are fulfilling user requirements and vendor service.

Table 22. Important Features as a Function of Plant Size

Number of Employees	Most Important Features
1 - 99	fault tolerance, relay-ladder language
100 - 999	fault tolerance, relay-ladder language
1000+	increased flexibility, standardization, flexible manufacturing

Table 23. Important Features as a Function of Period of Adoption

Period of Adoption	Most Important Features
prior to 1980	cost per channel, relay-ladder language
1980 - 1983	fault tolerance, relay-ladder, flexible manufacturing
1984 -	self diagnosis, fault tolerance

Figure 1

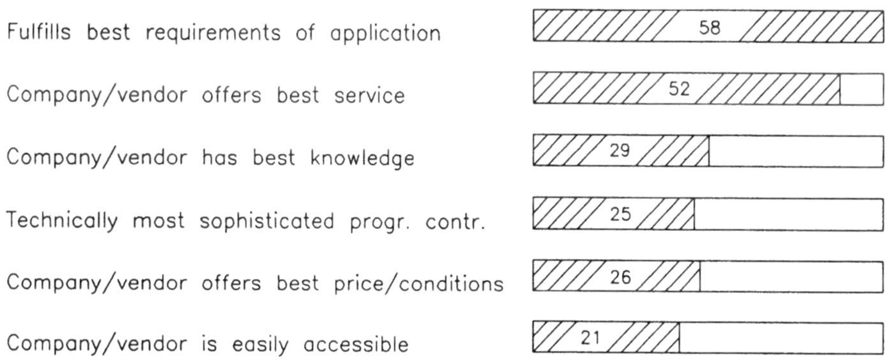

Reasons Influencing the Purchase of Programmable Controllers

11. REPORTED SUGGESTED IMPROVEMENTS

Even though the availability of a uniform protocol was not rated as very important for the use of programmable controllers, its improvement was suggested as the most important feature for more effective use in the future. Easier programmability with personal computers was the second most desired improvement, followed by the suggestion that more intelligence should be installed in programmable controllers. Cheaper prices were the fourth most desired improvement (Table 24).

Table 24. Reported Suggested Improvements

Improvement Index	Total Score
uniform protocol	157
easier programmability with personal computers	151
more intelligence	132
cheaper programmable controllers	130
rapid fault location modules (including peripherals)	130
cheaper associated equipment	128
less need for special skills	121
easier maintenance	117
greater versatility	113
better reliability	111
lower operating costs	96

Surprisingly, the price of the associated equipment was rated lower than the price of the programmable controller. Less need for special skills was also rated relatively low even though lack of technical in-house expertise was one of the major complaints.

12. CASE STUDY INTERVIEW

The case study was based on a Fortune 500 company in the food industry. The specific plant studied was founded in 1947 to supply the surrounding region with a variety of products. To increase product line flexibility, programmable controllers were installed in the 1970s as a substitute for relay boards. During the time the programmable controllers were in use, ideas for improving the processes were generated by various corporate groups. The flexibility of the programmable controller allowed these changes to be accomplished easily. Although a comparison between the choice of a programmable controller and its alternatives was made, a full feasibility study was not undertaken because the whole system was not considered.

Since the installation of the first programmable controller was done in the 1970s, exact data was difficult to obtain. The planning of the most recent system, including programmable controller and software, was performed by the vendor. The company is very satisfied with vendor service, maintenance and technical support. Software is easily available and, thus, it was given a low position in the questionnaire rating. Currently, the time needed to fully implement new programmable controllers is about one month.

The reliability of programmable controllers was not a major concern. Most of the problems experienced have been related to the associated equipment. In order to address these problems, technical expertise was found to be a necessity. This firm used in-house training for its personnel.

The firm uses programmable controllers for process control, sometimes with PID-loop (Proportional-Integral-Differential), and for material handling as a stand-alone device. Integrating the programmable controllers in an information network is planned. In the next stage it will provide production management with information regarding yield, and in the future it will provide control of Just In Time (JIT) systems. Measurements used by the firm for evaluation of the project are Internal Rate of Return as well as payback period.

13. CONCLUSION: POLICY IMPLICATIONS AND RECOMMENDATIONS

The survey demonstrates that the use of programmable controllers has been successful for a great majority of adopters. Programmable controllers are important as an indicator of the extent of the adoption of new production technology by industrial firms. A majority of the responding users indicated an intention to buy additional programmable controllers. The important features according to the users are relay ladder programmability and fault tolerance. Interestingly, the case study interview showed that the availability of software was rated as the least important feature due to its ready accessibility and quality. The period of time a programmable controller has been in operation can also change the emphasis on importance of attributes. For example, the importance shifted from self diagnosis and fault tolerance to relay ladder language and cost per channel as period of use proceeded.

Users of programmable controllers enjoyed numerous benefits in terms of time, throughput and quality. The most important performance indices according to programmable controller users were payback period and reduced repair and downtime. The advantages in using programmable controllers are: higher reliability, increased flexibility for production changes and the adaptability of programmable controller performance in diverse industrial activities.

Despite the overwhelming positive experience of users, certain limitations and disadvantages need to be considered:

(1) The main difficulties encountered by users include lack of proper planning of installation and integration, as well as the lack of technical in-house expertise. Adequate in-house expertise would be the best protection against these problems. Outside support services (e.g., programmable controller manufacturers/vendors) can also be extremely useful, but did not always prove satisfactory because of low quality customer service and lack of knowledge concerning individual company operational systems. An improvement could be achieved through more in-house and manufacturer training.

(2) Programmable controllers should not be used as isolated pieces of equipment, but as key elements in wider production systems using advanced manufacturing technology.

(3) Although programmable controllers are very reliable, the responding firm of the case study indicated that the system required an average of about three months before it operated reliably enough to be placed into commercial production. This lag time, however, appeared to stem more from associated equipment than programmable controllers.

(4) Programmable controllers tend to be used in large plants. Smaller plants may find it more difficult to introduce programmable controllers into their operational systems due to the lack of funding, expertise and awareness of advanced technology.

(5) Opposition within the plant can also be an obstacle. Surprisingly, major opposition in most firms arose mainly from the management and not from the shop floor.

(6) The high cost of equipment can be a potential impediment to the installation of programmable controllers, especially in regard to small plants. It should be noted, however, that this high cost includes the system as a whole and not just the programmable controller alone.

For the future, users are anticipating two main improvements: a uniform protocol, and easier programmability with personal computers. In addition, rapid fault location modules (including the peripherals), cheaper programmable controllers and greater intelligence are also desired features.

14. REFERENCES

Bulkeley, D. (1987). "Programmable Controllers: Workhorses for Designers," **Design News**, September 21, 78-80.

Bentley, J. M. (1987). "Interfacing Modern Drives with Mill Distributed Control Systems," **IEEE Transactions on Industry Applications, IA-23** (3), 392-397.

Hayner, A. M., L. R. Martin, P. P. Mishne, R. R. Schreiber, and R. N. Stauffer. (1988). "Forecast: Healthy PLC Market.," **Manufacturing Engineering**, November, 42-44.

IEEE Standard Dictionary of Electrical and Electronics Terms. (1977). (2nd ed.). New York: Wiley & Sons, 528.

Jacic, L. (1988). "Programmable Controllers in Water Treatment Plant," **Advances in Instrumentation, Proceedings of the Instrumentation Society of America '88 International Conference and Exhibit, 43** (1). Houston, Texas, 247-261.

Norris, D. M. and M. A. Kross. (1988). "Small Batch Control Projects," **Advances in Instrumentation, Proceedings of the Instrumentation Society of America '88 International Conference and Exhibit 43** (1). Houston, Texas: Instrumentation Society of

America, 19-35.

Raia, E. (1986). "The Other PC Boom - Programmable Controllers," **Purchasing 101**, August 21, 38-42.

Smith, S. E.(1988). "Programmable Controllers," **International Encyclopedia of Robotics: Applications and Automation**, Richard C. Dorf (Editor-in-Chief). New York: Wiley & Sons, 1187-1200.

Struger, O. J. (1985). "Programmable Controllers - Past and Future," **Proceedings of the Conference on Programmable Controllers '85 17th-19th July 1985**. London, UK: Peter Peregrinus, Ltd, 1-7.

Sugiyama, H. and W. J. Williams. (1988). "Open Distributed Control System Architecture for Total Factory Automation," **Advances in Instrumentation, Proceedings of the Instrumentation Society of America '88 International Conference and Exhibit 43** (3). Houston, Texas: Instrumentation Society of America, 1171-1177.

THE INTRA-FIRM DIFFUSION OF CADCAM IN THE U.K. MECHANICAL ENGINEERING INDUSTRY

M.J. Pokorny, V.G. Lintner, M.M. Woods and M.R. Blinkhorn
The Business School
The Polytechnic of North London
Holloway Road
London N7 8DB England

ABSTRACT

This paper presents an empirical analysis of the intra-firm diffusion of CADCAM in the UK mechanical engineering industry. We argue that it is only at this level of disaggregation that any detailed insights into the diffusion process can be derived. Diffusion process is defined as a two stage process, involving the initial decision of whether or not to adopt, followed by the decision relating to the rate at which adoption is to take place. A distinction is drawn between the adoption of pure manufacturing technology (essentially the adoption of CNC machinery), and the adoption of design technology (CAD). The empirical results of the paper are based on a sample survey of 100 mechanical engineering establishments, and separate models are specified and estimated for the adoption of CNC and the adoption of CAD. The main finding of the analysis is that the major impediment to adoption can be characterized as demand-side in nature, with supply-side influences being relatively unimportant. A further implication of the findings is that firms which place an over-reliance on the results of formal investment appraisal analyses tend to be somewhat conservative in their approach to the adoption of advanced manufacturing technologies, and it is those firms which are flexible in the appraisal process which tend to be the innovators.

1. INTRODUCTION

The empirical literature associated with the economic analysis of technical change has been dominated with analyses of either inter-industry diffusion or inter-firm diffusion. The reason for this emphasis is simply the result of data availability at the industry rather than at the firm level. The result is that empirical conclusions that have been reached about the adoption of new technology are relatively general, with only limited insights available into the specific nature of the adoption process. These general conclusions contrast with the much more detailed ones found in the theoretical literature.

Nonetheless, if detailed insights are required into the nature of the process by which advanced manufacturing technology is adopted, these can only be obtained by examining the adoption process at the level of the firm. Detailed empirical analysis can only proceed once a sufficiently specific form of advanced technology has been identified.

This paper represents an empirical contribution to the analysis of technological change, and is based on a cross-sectional study of the behavior of a sample of establishments. The objectives of this paper are necessarily modest, with only a very specific form of advanced manufacturing technology--Computer Aided Design/Computer Aided Manufacture (CADCAM)--within a single industry investigated. Thus, although the number of general inferences about technological change that can be drawn from such an analysis are limited, the approach does allow for a relatively detailed analysis of the adoption process.

The adoption of CADCAM is only considered within a single industry--mechanical engineering in the United Kingdom (U.K.), in order to limit the need for extensive ceteris paribus controls. CADCAM was selected for analysis since it represents a highly advanced form of microelectronic technology, and therefore offers the possibility of drawing wider inferences about the adoption of microelectronics in general. Mechanical engineering was selected since it was once a major industry in the U.K., although it suffered conspicuously during the recession of the late 1970s and 1980s, and because it is an obvious potential user of CADCAM. The data upon which the results of this paper are based were derived from a sample survey of 100 English mechanical engineering establishments, conducted during the summer of 1983.

Computer Aided Design (CAD) can be broadly defined as the

application of computer technology to the process of product design. At its most advanced level, CAD combines highly developed computerized graphical routines with the facility for direct interaction with these routines so that the designer can vary and develop the product design. However, many lower level CAD systems are available, including computerized drafting systems, which do not strictly offer the facility for direct design variation, but instead provide a means for efficiently and accurately producing and amending blueprints for existing designs. At the lowest level, a CAD system would involve some form of pre-programmed routine for design/engineering calculations, but would not involve any computerized graphics.

Computer Aided Manufacture (CAM) covers a wide range of levels of sophistication. At its most general level, CAM would be defined as involving some form of pre-programmed machinery for use in the manufacturing process. The most common form of pre-programmed machinery is Computer Numerically Controlled (CNC) machine tools, which can be programmed to perform some predetermined manufacturing task. At the advanced level, fully integrated CADCAM systems are available, which allow for product designs developed on a CAD system to be transferred directly to computerized machine tools, so that the final product can be manufactured without physical intervention on the part of any operative.

The use of such advanced systems is relatively rare in the U.K., and typically the CAD systems and CNC equipment are physically separated. CNC is more widely used than CAD, and it is common for firms just to use CNC equipment and to use more traditional methods for product design, particularly when a firm is producing a product range requiring only limited design variation from year to year. Alternatively, the full automation of design and manufacturing activities will often be undertaken iteratively, with automation of the manufacturing process generally preceding that of the design process.

In our discussion, we will first briefly interpret the nature of the intra-firm adoption process, and describe how this process might be modelled. In particular, we will identify the major influences on the adoption process. The following section will present the empirical results, and we will then conclude with a summary and discussion of the results.

2. MODELLING THE INTRA-FIRM DIFFUSION PROCESS

The intra-firm diffusion of a technical innovation can be most usefully viewed as a two-stage process. The first stage involves the initial decision to adopt the technology. The next stage involves the decision about the rate at which the technology is to be adopted. Our objective is to develop models of these two sequential decisions.

2.1 The Initial Adoption Decision

We are first concerned with identifying those factors that would induce a non-adopting establishment in time period t-m to either adopt or remain a non-adopter by period t. Thus, let A_i be the dependent variable of interest, where A_i takes on the value 1 if establishment i adopts during the period t-m to t, and the value 0 if establishment i remains a non-adopter. In specifying a model that explains the variation in A across a sample of establishments, we are deriving a model that will predict the probability of an establishment adopting the technology.

We could characterize A as being a function of three distinct sets of influences--**technical, supply-side** and **demand-side. Technical influences** are those factors that relate to how suitable the establishment's output or mode of operation is for the adoption of new technology. Such influences are exogenous in the sense that, at a point in time, such factors are not susceptible to any form of policy intervention. CNC technology establishments, which are most likely to adopt, ceteris paribus, generally have a specialized or limited product range that involves repetitive production and long production runs. In terms of design technology, the greater the extent of design input into the production process, the greater will be the need for CAD, and hence the higher the probability of adoption.

In terms of **supply-side influences**, a number of elements can be isolated. First of all, certain factors might "push" an establishment into adoption. Establishments operating in some sense inefficiently, or in particular, those suffering from declining efficiency levels, might be induced to reappraise their production methods, and consider the adoption of more advanced manufacturing technology. This situation commonly occurs when the establishment is operating in an environment in which the use of advanced technology is prevalent among its competitors, leading it to perceive outmoded production methods as the cause of operational inefficiency.

Although its role is ambiguous, a second supply-side factor is establishment size. Larger establishments may tend to be involved in relatively long production runs, and thus be more suited to the use of advanced technology, at least at the manufacturing level. However, such influences have already been accounted for under the above described technical influences. Thus, further justifications must be found for the inclusion of size. Certainly, we would expect larger establishments to adopt a higher absolute **level** of advanced technology, ceteris paribus, but there is no reason to expect that size will be related to the **probability** of adoption. Alternatively, it might be argued that larger establishments are in some sense more sophisticated and hence more receptive to innovation. Larger establishments might also have easier access to finance, which would facilitate adoption. However, it might alternatively be argued that larger establishments may experience relatively greater organizational upheaval as a result of adoption, making it conceivable that size may to some extent act as an impediment. Thus, since size acts as an imperfect proxy for a number of diverse influences, its role is an ambiguous and unpredictable one. A final supply-side influence that has been argued to be relevant is the role played by trade unions. Given the labor-saving nature of many forms of new technology and the perceived de-skilling consequences of its adoption, it is expected that trade unions will act as an impediment.

Finally, we can consider the **demand-side influences** on the probability of adoption. Two dimensions of the role played by demand can be identified. First of all, the higher the level of demand for the establishment's output is in time period t-m, the higher is the probability of adoption over the ensuing periods. Conversely, low demand would inhibit adoption since it may be difficult to justify the investment and adjustment required during a period of depressed economic activity. A second influence on the demand-side is the role played by expectations. Optimistic expectations about future demand levels, given the level of demand in time period t-m, would be expected to stimulate adoption and pessimistic expectations to inhibit it.

Thus, our adoption model can be summarized as follows:

$$A_i = f(SUIT_i, EFF_i, SIZE_i, TU_i, DEM_i, DEM^e{}_i) \qquad (1)$$

where A_i is the dependent variable of interest, $SUIT_i$ reflects the suitability of establishment i's output to new technology in time period t-m, EFF_i is a measure of efficiency, $SIZE_i$ is establishment size, TU_i reflects the influence of trade unions, DEM_i is the level of demand for establishment i's output in period t-m, and $DEM^e{}_i$ is a measure of expectations held in time period t-m about future demand levels.

2.2 The Rate of Adoption of New Technology

In analyzing the rate of adoption of a technical innovation, the common approach has been to treat the process of adoption as analogous to the spread of a contagious disease. Such an approach then implies that the level of usage of the innovation can be interpreted as following some form of sigmoid time path. We will take this approach here. Two specific forms of the sigmoid curve that have been found to be empirically useful are the logistic and Gompertz formulations. Thus, assuming a logistic formulation, an establishment's time path of adoption would be described by:

$$Y_t = \frac{k}{1 + e^{\alpha+\beta t}} \qquad (2)$$

An assumption of Gompertz adjustment would imply:

$$Y_t = k\,\alpha^{\beta^t} \tag{3}$$

where Y_t is the level of usage of the innovation in time period t, and k is the theoretical maximum, or optimum, level of usage--the level to which the establishment can be interpreted as adjusting. α and β, then, are the parameters that specify the nature of the adoption path.

From the logistic assumption, we can deduce from (2) that:

$$\frac{dY_t}{dt} = \frac{-\beta}{k}\; Y_t(k - Y_t) \tag{4}$$

and from the Gompertz assumption, we can deduce from (3) that:

$$\frac{dY_t}{dt} = -\ln\beta \;\; Y_t(\ln k - \ln Y_t) \tag{5}$$

Thus, in terms of the logistic formulation, the coefficient $-\beta/k$ measures the responsiveness of the establishment to the divergence between its optimum and actual level of usage. It therefore can be interpreted as reflecting the rate of adoption. Similarly, the rate of adoption from an assumption of the Gompertz adjustment would be reflected in $-\ln\beta$. Our approach here will be to derive an estimate of these coefficients for each of the establishments in the sample and then to specify a model that explains the variation in these coefficients. Thus, explaining the cross-section variation in these adoption rates should allow for the identification of those factors that influence the adoption process.

We are certainly not suggesting that an establishment remains on a fixed adoption path over time (Metcalfe, 1981). Clearly, a range of factors will induce an establishment to alter its behavior over time, and it would be of some interest to identify such factors. For example, the success of government microeconomic policies designed to stimulate adoption could be judged in terms of the extent to which firms have been induced to move on to higher adoption paths. Our approach attempts to explain why, at a given point in time, adoption behavior varies in the cross-section. This approach should then provide a basis for inferring how higher rates of adoption might be stimulated.

In order to estimate the logistic and Gompertz adoption rates, we require an estimate of the parameter β in both functions. First of all, consider the logistic formulation. Let Y_0 be the initial, or instantaneous, level of usage, once the decision to adopt has been made. Therefore, from (2) we have:

$$\alpha = \ln(k-Y_0) - \ln Y_0 \tag{6}$$

Substituting (6) into (2) and solving for β produces:

$$\beta = (1/t)\ \ln\{[Y_0(k-Y_t)]/[Y_t(k-Y_0)]\} \tag{7}$$

Thus, dividing (7) throughout by -k produces the expression for the logistic adoption coefficient.

Similarly, we can deduce from (3) that:

$$-\ln\beta = (-1/t)\ \ln\{\ln(Y_t/k)/\ln(Y_0/k)\} \tag{8}$$

and, therefore, (8) is the expression for the Gompertz adoption coefficient.

We will now consider the potential influences upon these adoption coefficients. As in the case of the initial adoption decision, we will categorize the influences into technical, supply-side and demand-side influences.

In terms of technical influences, the following factors can be identified:

The suitability of an establishment's output to advanced manufacturing technology. Again, we would expect this factor to be of relevance, and in particular, to exert a positive influence on the rate of adoption. Thus, not only would suitability increase the probability of adoption, but also it would allow for a more rapid rate of adoption, ceteris paribus.

An establishment's experience of using advanced technology. This factor is commonly mentioned in the literature (Mansfield, 1968). The argument is that the more familiar the establishment is with the innovation, the more confident the establishment will be about adoption, leading the establishment to adopt advanced technology more rapidly.

The length of time that the establishment has been using the technology. We would expect this variable to exert a negative influence on the rate of adoption for two reasons. First of all, the more recent the adoption of technology, the greater will be the pressure on the establishment to "catch up" to the existing users in the industry. Secondly, more recent adopters will be adopting in an environment in which more is known about the technology, putting them in a stronger position to adopt rapidly.

In terms of the supply-side influences, the same factors that were identified as being relevant in relation to the initial adoption decision would also be expected to influence the rate of adoption. Thus, operational inefficiency might be expected to stimulate adoption and thus imply more rapid adoption. While size has an ambiguous influence, trade unions might be expected not only to impede the initial adoption of new technology but also its subsequent rate of adoption.

On the demand side, only expectations may be expected to positively influence the rate of adoption. While a high current level of demand might be expected to induce an establishment to adopt, it is only expectations of future demand levels that will influence the rate of adoption. By definition, the decision as to the current rate of adoption is made with reference to expectations of the future, with the current level of demand being treated as, in effect, a bygone. In other words, the current level of demand may justify adoption, but it is only the nature of expectations about the future that can justify the rate at which adoption takes place.

Thus, in summary, the model explaining the rate of adoption can be expressed as follows:

$$RATE_i = f(SUIT_i, EXPER_i, USE_i, EFF_i, SIZE_i, TU_i, DEM^e{}_i) \qquad (9)$$

where $RATE_i$ is the rate at which establishment i is currently adopting, $EXPER_i$ and USE_i are establishments i's experience of the technology and the length of time the technology has been used, respectively, and the remaining variables are as defined previously.

3. EMPIRICAL RESULTS

The sample of 100 mechanical engineering establishments that generated the data to estimate the models specified in the previous section was selected from a sampling frame containing 3,145 establishments throughout Great Britain.[1] The sample was stratified regionally, but was not selected randomly. In order to obtain sufficient observations on users of CADCAM, particularly CAD, it was necessary to bias the sample toward establishments which, a priori, appeared to be likely users of CADCAM. The sampling frame indicated whether or not each establishment used a computer and also indicated the size of each establishment in terms of number of employees. Likely users of CADCAM were then defined as relatively large establishments that used a computer. Although this definition proved to be only a rough and ready guide, it did ensure that sufficient numbers of users were interviewed. A second regionally stratified random sample was also selected in order to estimate a number of broad population characteristics, particularly the extent of adoption within the industry. This second sample consisted of 200 stratified and randomly selected establishments, of which 54 completed mail questionnaires. The results presented here are derived from the larger sample of 100 establishments, from which more detailed data were collected. Employees of the establishments were personally interviewed in the summer of 1983, and data were collected relating to each of the years 1980, 1981, 1982 and for the period up to the date of the interview.

Forty-four establishments used some form of advanced manufacturing technology in 1980, and this number increased to 68 by mid-1983. In 1980, 17 establishments used CAD and 37 used CNC.

---

[1]Supplied by Market Location Limited.

By 1983, 43 used CAD, and 55 used CNC. Average establishment turnover in 1982 was £11.427m, with a coefficient of variation of 0.57, and the average number of full-time employees at the time of the survey was 486, with a coefficient of variation of 0.48. For comparative purposes, the corresponding figures for the smaller random sample of 54 establishments were as follows: 12 establishments were using some form of advanced technology in 1980, growing to 22 by 1983. All of these establishments used CNC machinery, but only one was using CAD in 1980, with this number increasing to five by 1983. Average establishment turnover in 1982 was £1.751m, with a coefficient of variation of 0.62. Average full-time employees per establishment was 105, with a coefficient of variation of 0.48.

Given the manner in which CADCAM technology is adopted, and, in particular, the distinction that can be drawn between manufacturing and design technology, we considered it appropriate to treat the adoption of CNC and CAD as separate decisions. Thus, separate models will be estimated for CNC and CAD.

3.1 The Initial Adoption of CNC

Since most of the data collected related to the period 1980 to 1983, only adoption behavior that occurred during this period can be examined in any detail. Therefore, we could only consider those establishments that were non-users in 1980. The dependent variable thus takes on the value 1 if an establishment adopted CNC during this period, and the value 0 if the establishment remained a non-user. The various independent variables were measured as follows:

The suitability of the establishment's output to CNC. We have argued that, in the case of CNC, a number of dimensions to this explanatory factor exist, namely the degree to which the establishment specializes, the extent of repetitive production and the length of production runs. However, the sample data only allowed for the derivation of a measure of the extent of specialization, thus restricting the testing of the relevance of output suitability. The measure used is the proportion of the establishment's output which was accounted for by the most important product in 1982. The justification for using the degree of specialization in 1982 (rather than 1980) is that the adoption decision in 1980 will be a function of the expected suitability of the establishment's output, rather than current suitability--the decisions relating to output structure and adoption could in this sense be interpreted as joint ones. In any event, the correlation between output specialization in 1980 and 1982 for the 51 non-user establishments in 1980 for which observations were available was 0.96. Thus, the use of either variable makes little difference, but the advantage of use of the 1982 variable is that it provides more sample observations. We denote this chosen variable by SPEC.

Efficiency. We based the measure of efficiency on the profitability of the establishment, as reflected by the ratio of pre-tax profits to turnover. The variable used was the simple change in profitability between the two years at the beginning of the adoption period (1980 and 1981). Thus, establishments for which this variable is negative could be interpreted as suffering from declining efficiency levels, which might induce them to consider the adoption of alternative production methods. Conversely, establishments for which this variable is positive or zero could be interpreted as maintaining or even improving their productive efficiency, which makes them less likely to be under the same pressure to re-evaluate their production methods.

Establishment size. Since CNC is a manufacturing technology, the appropriate size variable is the scale of the establishment's manufacturing activity. The variable used to represent this activity is the number of blue-collar employees in 1980. For estimation purposes, the natural logarithm of this variable was used, which is denoted by LBC80.

Trade unions. Given that CNC is a shop-floor based technology, any trade union pressure would be expected to derive primarily from blue-collar unions. The variable used to represent trade union pressure was based on the percentage of the blue-collar workforce unionized in 1980, 1981 and 1982. A number of potential measures can be derived from these data, depending on whether it is considered appropriate to assume that unions exert a continuous or discrete influence. If the nature of union pressure is assumed to be continuous, then a potential measure of union strength is simply union density in 1980. If union pressure is assumed to be reflected as some form of discrete measure, a binary variable is used. In this case, a series of variables were used: a variable set to 1 if union density in 1980 exceeded 40 percent, 0 otherwise; a second variable set to 1 if union density exceeded 50 percent, 0 otherwise; and a final variable set to 1 if density exceeded 70 percent, 0 otherwise.

In all cases (discrete or continuous), the statistical results were virtually identical. Thus, for illustrative purposes, a binary variable that is set to 1 if union density exceeds 50 percent will be used. This variable will be denoted by BCTU80. Certainly, it can be argued that such a simple measure cannot possibly capture the variety of forms that union influence might take. However, the survey collected a range of attitudinal information relating to management's perceptions of union influence, and these data do not conflict with the conclusions that are reached on the basis of the measure used here.[2]

Demand. Data was collected about the proportion of the establishment's total capacity utilized in 1980, 1981 and 1982. The demand variable used is the level of capacity utilization in 1980.

Expectations. As is the case in most empirical studies, no ideal measure of expectations is available, and the measure used here is in large part conditioned by the nature of the sample data. The simplest assumption that can be made is that expectations of future demand levels is some function of the trend in past demand levels. However, given that the sample data have only a limited time-series dimension (1980 to 1982), we are restricted to the use of a relatively simple expectations variable. The variable used is the simple change in capacity utilization between 1980 and 1982. Since we are therefore using the <u>actual</u> change in capacity utilization as a measure of the expected change over the adoption period, such a measure can only be justified if expectations can

---

[2]See Lintner, et al. (1987) for a more detailed discussion of these issues.

be assumed to be "rational."[3]

3.2 The Initial Adoption of CAD

The dependent variable here takes on the value 1 if an establishment adopted CAD during the period 1980 to 1983, and the value 0 if the establishment remained a non-user. In terms of the independent variables, efficiency, demand and expectations were measured the same way as CNC. Given that CAD is an office-based technology, any trade union influence would be expected to derive from white-collar unions. Thus, the union variable used was a binary variable taking on the value 1 if union density exceeded 50 percent in 1980, and 0 otherwise. The remaining independent variables were defined as follows:

The suitability of the establishment's output to CAD. The appropriate variable here is a measure of the extent to which the establishment is involved in product design. We assumed that the scale of an establishment's design activity was directly proportional to the numbers employed in the categories of draftsmen, white-collar engineers and other scientists/ technologists at the beginning of the adoption period (1980). Thus, the larger the number of such employees, the greater is the extent of design activity, and hence the greater the probability of adoption. Those establishments that have no employees in these categories are assumed not to be engaged in design activity, thus having no need for CAD. Such establishments are excluded from further analysis. Often, the product design for such establishments takes place in other divisions within the firm or is done by the parent company. Thus, these establishments' non-adoption can be explained without recourse to the range of factors identified previously. For purposes of estimation, the natural logarithm of the suitability variable, denoted by LDES80, was used.

---

[3]However, see Tompkinson and Common (1983) where the validity of such an assumption has been questioned.

Establishment size. In effect, we have already included a size variable in the model via the inclusion of LDES80. However, to the extent to which this variable is not a perfect proxy for establishment size, and thus might not fully reflect the range of additional influences that could be hypothesized to derive from size, a further variable will be included. The variable used, denoted by LWF80, is the size of the establishment's full-time workforce in 1980 in logarithmic form.

We will estimate the models explaining the adoption of CNC and CAD in logit form. Table 1 presents various specifications of the logit equation for the adoption of CNC, and Table 2 presents variations on the CAD equation. The varying sample sizes associated with the equations result from differing degrees of missing observations among the independent variables, particularly the profitability data used to construct the efficiency variable.

Model 1 in Tables 1 and 2 is the estimate of the fully specified CNC and CAD equation, respectively, as summarized in equation (1). The remaining models in both tables represent variations on this basic specification.

In terms of the CNC equations in Table 1, the most significant influence on the adoption decision would appear to be establishment efficiency, which is consistent with the hypothesis that inefficiency induces adoption. A weak influence is exerted by the level of demand, and, if anything, unions could be interpreted as exerting a positive, although weakly significant, influence on adoption. Establishment size, expectations and degree of specialization are all insignificant, and thus Models 2, 3 and 4 successively omit these variables.

From Table 2, the adoption of CAD would appear to be largely influenced by demand-side factors, with both the level of demand and expectations exerting a positive influence. None of the supply-side variables are significant, and thus Models 2, 3 and 4 successively omit these variables. As expected, establishments with a high level of design input are more likely to adopt.

These results can be rationalized by arguing that, since CNC is a relatively well-established technology, inefficient non-adopters are more likely to attribute their inefficiency to out-moded production methods. The competitive pressures to adopt

Table 1. Logit Equations for the Adoption of CNC

VARIABLE	MODEL 1	MODEL 2	MODEL 3	MODEL 4
Technical Influences				
SPEC	-0.030	-0.030	-0.024	
	(1.01)	(1.07)	(0.95)	
Supply-side Influences				
EFF	-31.270	-31.069	-24.990	-22.888
	(2.08)	(2.20)	(2.41)	(2.34)
LBC80	-0.188			
	(0.22)			
BCTU80	2.182	1.679	1.602	
	(1.29)	(1.59)	(1.44)	(1.44)
Demand-side Influences				
DEM	0.067	0.066	0.057	0.074
	(1.48)	(1.48)	(1.48)	(1.84)
DEM^e	0.057	0.057		
	(0.96)	(0.99)		
Constant	-5.574	-6.247	-5.577	-8.911
-2 log (likelihood ratio)	16.71	17.22	16.17	15.71
Sample Size	31	32	32	35
Number of observations correctly predicted by the model	27	28	28	32
Number of adopters	9	9	9	9

Notes (1) The absolute value of conventionally calculated t-statistics are shown in brackets.

(2) -2 log (likelihood ratio) is the test statistic derived from a likelihood ratio test of the null hypothesis that the equation is not superior to an equation containing just a constant term. The test statistic has an approximate chi-square distribution with degrees of freedom equal to the number of independent variables in the equation, excluding the constant term.

Table 2. Logit Equations for the Adoption of CAD

VARIABLE	MODEL 1	MODEL 2	MODEL 3	MODEL 4
Technical Influences				
LDES80	0.643 (1.30)	0.817 (2.32)	0.801 (2.72)	0.821 (3.06)
Supply-side Influences				
EFF	-1.588 (0.28)	-1.532 (0.27)		
LWF80	0.359 (0.49)			
WCTU80	0.357 (0.35)	0.456 (0.46)	0.587 (0.73)	
Demand-side Influences				
DEM	0.032 (1.88)	0.032 (1.86)	0.025 (1.66)	0.014 (1.06)
DEM^e	0.066 (1.93)	0.062 (1.87)	0.067 (2.17)	0.054 (2.03)
Constant	-6.564	-5.042	-4.750	-3.668
-2 log (likelihood ratio)	13.13	12.88	16.12	15.11
Sample Size	43	43	55	58
Number of observations correctly predicted by the model	35	34	43	46
Number of adopters	18	18	22	25

Notes (1) The absolute value of conventionally calculated t-statistics are shown in brackets.

(2) -2 log (likelihood ratio) is the test statistic derived from a likelihood ratio test of the null hypothesis that the equation is not superior to an equation containing just a constant term. The test statistic has an approximate chi-square distribution with degrees of freedom equal to the number of independent variables in the equation, excluding the constant term.

are more intense when the use of the technology is more widespread, and the adoption models in Table 1 can therefore be interpreted as reflecting the imitative behavior of non-adopters. Such a rationalization might also explain the insignificance of establishment size on adoption. The initial adopters of CNC in the industry may well have been the larger establishments, but as diffusion throughout the industry proceeds, it is the smaller establishments that tend to remain non-adopters, leading the significance of size on adoption to die out over time.

On the other hand, CAD is a newer technology, and therefore greater uncertainty surrounds its use. Since the use of CAD is less widespread than that of CNC, the same competitive pressures are not associated with its adoption. Thus, we might expect conservatism to play a greater role in the adoption decision. Only those establishments that are in a relatively strong position with respect to the demand for their output might have the confidence to adopt, and this is how the results of Table 2 can be interpreted. To the extent to which LDES80 is a reflection of establishment size, we could also interpret size as having a positive influence on the adoption of CAD. However, it might be argued that the efficiency variable used in Table 2 is inappropriate in the case of CAD--that what is required is strictly a measure of **design** efficiency. However, no such measure was available from the sample data.

Further qualifications that must be made in the interpretation of the estimated models in Tables 1 and 2 concern the potential influence of additional factors that could not be measured from the sample data. Some of these have already been mentioned in the discussion of the role of establishment size. Access to finance, which is expected to have an influence on adoption, could vary among the firms. The flexibility of management in evaluating the costs and benefits of new technology might be relevant, in the sense that the rigid interpretation of the results of traditional investment appraisal methods might be expected to inhibit adoption. Establishments that are currently going through a replacement cycle may be more likely to consider adoption since advanced technology might be considered more profitable for those establishments with older capital equipment. However, we would expect such a replacement cycle influence to be at least partially reflected in our efficiency measure.

3.3 The Rate of Adoption of CNC

Respondents who used CNC were asked to indicate the proportion of their output that is currently produced by CNC and the number of years that the establishment has been using CNC. Thus, in terms of the expressions for the logistic and Gompertz adoption rates in (7) and (8) above, let t denote the number of years CNC has been used, and Y_t the proportion of output currently produced by CNC. Observations on Y_0 and k are required in order to derive estimates of the adoption rates.

We will set Y_0, the initial level of usage of CNC, to 0.01 for each user establishment. Thus, we interpret each establishment as adopting from the same (instantaneous) base level. However, a more fundamental problem arises in the measurement of k, the establishment's optimum level of usage of CNC. It could be argued that this optimum level will vary across establishments, and will be a function of certain technical characteristics relating to the nature of the establishment's output. However, since the sample data did not allow for the derivation of a satisfactory measure of k for each establishment, the approach taken was to set k equal to its theoretical maximum of 1 for all observations. Thus, we interpret each establishment as adjusting toward 100 percent utilization of CNC, and the measures derived from (7) and (8) reflect the rate at which this was being achieved in 1983. Some of the independent variables in the model will then be interpreted as accounting for the extent to which it might not be reasonable to assume the possibility of 100 percent utilization.

In terms of the independent variables in (9), $SUIT_i$, EFF_i and DEM_i^e are the same variables used in the CNC adoption equation, in which SUIT was measured by SPEC. Thus, in terms of the measurement of expectations, we would have to argue that the past trend in capacity utilization (1980 to 1982) reasonably reflects expectations about the future. Establishment size is measured by the logarithm of the size of the blue-collar workforce in 1982 (denoted by LBC82), and the trade union variable (denoted by BCTU) is a binary variable, taking on the value 1 if average blue-collar density over the years 1980 to 1982 exceeded 50 percent, the value 0 otherwise. USE_i is simply the number of years establishment i has been using CNC, entered in logarithmic form. We measure $EXPER_i$, establishment i's experience of CNC, by the proportion of its output that is produced by CNC. However, it would be inappropriate to use the establishment's current level of usage. Rather, since we would expect a lagged response from experience to the current rate of adoption, the variable used is the proportion of the establishment's output produced by CNC in 1980, the only other year for which this information was collected.

Note, then, the way in which SPEC must be interpreted in this formulation. We would expect that the more the establishment's output is suited to CNC, the greater will be the need for CNC, ceteris paribus, and hence the higher will be its rate of adoption. However, the larger is SPEC, the greater is the scope for a higher rate of utilization of CNC, and hence the closer will be the establishment's optimum level of utilization to 100 percent. Conversely, establishments with a diverse product range are likely to have an optimum level of utilization somewhat below 100 percent, and thus their **estimated** rate of adoption from (7) and (8) would understate their actual rate of adoption relative to their internally determined optimum. In other words, low observed adoption rates will be associated with low values of SPEC, reinforcing the expectation of a positive association between these variables. Similarly, it might be argued that, the longer the establishment has been using CNC, ceteris paribus, the closer the establishment will be to its own optimum. This implies that the rate of adoption as estimated here will be more of an understatement the longer that CNC been used, further implying a negative association between USE and RATE.

In Tables 3 and 4, we present least squares estimates of various specifications of (9) for the adoption of CNC, assuming logistic and Gompertz adoption, respectively. Although the conclusions that can be drawn from Tables 3 and 4 are virtually identical, the logistic equations produce higher R^2s, perhaps implying that an assumption of logistic adjustment is the more

Table 3. Least Squares Regressions Explaining Logistic Adoption Rates for CNC

VARIABLE	MODEL 1	MODEL 2	MODEL 3	MODEL 4
<u>Technical Influences</u>				
SPEC	0.006	0.006	0.006	0.003
	(1.52)	(1.77)	(1.77)	(1.29)
EXPER	0.014	0.014	0.013	0.009
	(2.44)	(2.60)	(2.81)	(2.38)
USE	-1.445	-1.472	-1.456	-1.308
	(6.72)	(7.28)	(7.66)	(9.06)
<u>Supply-side Influences</u>				
EFF	0.960	1.460	1.574	
	(0.71)	(1.38)	(1.63)	
LBC82	-0.006			
	(0.08)			
BCTU	-0.083	0.113		
	(0.37)	(0.55)		
<u>Demand-side Influences</u>				
DEM^e	0.022	0.020	0.019	0.015
	(3.31)	(3.91)	(4.06)	(3.48)
Constant	2.994	3.020	2.953	2.881
	(4.75)	(6.87)	(7.27)	(9.76)
$\bar{R}^2$	0.744	0.755	0.760	0.711
Sample Size	26	27	29	41

Notes (1) The absolute values of conventionally calculated t-statistics are shown in brackets.

Table 4. Least Squares Regressions Explaining Gompertz Adoption Rates for CNC

VARIABLE	MODEL 1	MODEL 2	MODEL 3	MODEL 4
Technical Influences				
SPEC	0.003	0.003	0.002	0.001
	(1.60)	(1.85)	(1.84)	(1.20)
EXPER	0.007	0.008	0.007	0.006
	(3.91)	(3.18)	(3.62)	(3.24)
USE	-0.495	-0.503	-0.499	-0.461
	(5.22)	(5.67)	(6.02)	(7.16)
Supply-side Influences				
EFF	0.615	0.717	0.775	
	(1.04)	(1.54)	(1.84)	
LBC82	-0.009			
	(0.23)			
BCTU	-0.043	-0.053		
	(0.44)	(0.58)		
Demand-side Influences				
DEM^e	0.008	0.008	0.008	0.005
	(2.79)	(3.55)	(3.71)	(2.80)
Constant	1.012	0.980	0.955	0.973
	(3.63)	(5.09)	(5.40)	(7.40)
$\bar{R}^2$	0.630	0.651	0.664	0.585
Sample Size	26	27	29	41

Notes (1) The absolute values of conventionally calculated t-statistics are shown in brackets.

reasonable one.

Thus, from the full specification of Model 1 in both tables, we would conclude that the technical influences on adoption are significant (with SPEC being only marginally so), and of the expected direction. Expectations exert a positive and highly significant influence on the rate of adoption, but no role appears to be played by supply-side factors. In Models 2 and 3, establishment size and union influence are successively omitted, the only consequence being that efficiency now appears to exert a marginally significant and positive influence on the adoption rate. Thus, while inefficient production might induce adoption, it would appear that, if anything, once adoption has taken place, it is only improving profitability that stimulates further adoption--an increased rate of adoption can only be justified within the context of the benefits being realized. Model 4 in Tables 3 and 4, in which EFF is finally omitted, is included simply to allow for the examination of the remaining variables within the larger sample that results. Thus, the broad conclusions drawn above still hold, particularly with respect to the influence of expectations, which remains positive and highly significant one.

3.4 The Rate of Adoption of CAD

Unfortunately, the sample data did not allow for a sufficiently detailed analysis of the factors that influence the rate of CAD adoption. The major problem was in simply deriving an adequate measure of the level of usage of CAD. Data was collected relating to the value of the establishment's CAD systems, and thus measures were derived on the basis of the value of CAD per capita. However, none of these measures proved satisfactory. Data relating to the current level of capacity utilization of the establishment's CAD systems was also collected, and rate measures analogous to those derived for CNC were calculated. The difficulty with such measures is that capacity utilization will also reflect demand influences, not just the speed with which the establishment is incorporating CAD into its operations. In any event, because of missing data, relatively few sample observations were available using such an approach, and those that were available produced insignificant statistical results.

However, perhaps a more fundamental conceptual problem in analyzing the rate of adoption of CAD exists. Once an establishment has decided to adopt, the selection of an appropriate CAD system will be highly specific to the establishment's needs, particularly with respect to the nature of the design activity in which it is involved. Therefore, the concept of a continuous rate of adoption over time is perhaps a misleading one--the purchase of a CAD system is in the nature of an "all-or-nothing" decision, the crucial decision being the initial one of whether or not to adopt. Certainly, it may be of some interest to examine the factors that influence the specific form of CAD system that is finally purchased. It may also be of some interest to examine the rate at which a **given** CAD system is absorbed into the design and production process. However, the relevant influences on these decisions are likely to be purely technical ones and highly establishment specific, in contrast to the range of influences with which this paper has been concerned.

4. SUMMARY AND CONCLUSIONS

In this paper, we have been concerned with the empirical analysis of the factors that influence the adoption of CADCAM in the U.K. mechanical engineering industry. Even though the focus of this paper has been limited, considerable compromises still had to be made in using survey data to examine empirically the complex process of the adoption of advanced manufacturing technology. In particular, the survey data did not allow for the examination of the range of behavioral and informal influences that might be expected to play some role in the adoption process. Therefore, the conclusions that can be drawn from our analysis should be interpreted, in this sense, as partial.

In the case of CAM, and in particular, CNC, we characterized the decision as being a two-stage one, involving the initial decision of whether or not to adopt, followed by the decision as to the rate at which adoption is to take place. We categorized the potential influences on these decisions into technical, supply-side and demand-side factors.

In terms of the initial adoption of CNC, the major influence came from the supply-side, in which it seemed that declining profit performance "pushed" non-users into adoption. We could interpret the significance of this influence as providing evidence that some firms have been forced into re-evaluating their production methods and pushed into emulating their more successful--and technologically superior--competitors. It follows, therefore, that such firms must evaluate the adoption of CNC somewhat informally, and not rely overly on the results of a formal investment appraisal analysis. Indeed, it might be argued that formal investment appraisal methods are quite inappropriate within the context of evaluating the adoption of an advanced or new technology. While the costs of adopting such a technology are straightforward, the expected benefits are much more difficult to measure, particularly because of the radical and often unexpected changes in production methods that result. Certainly evidence was available from our survey, albeit of a somewhat anecdotal and informal nature, implying that successfully adopting firms typically did not undertake detailed investment appraisals, but relied more on a technical appreciation of the capabilities of the technology, and how this technology might be incorporated into the firm's existing production methods. The conflict between the requirements of the accountant and the engineer was a common one, and those firms within which the engineer's influence was the greatest tended to be the more innovative adopters.

On the demand-side, a marginally significant and positive influence was exerted by the level of demand. The major influences on the rate of adoption of CNC were expectations about future demand levels and about technical factors. The sample data only allowed for an analysis of the decision relating to the initial adoption of CAD, although it was suggested that the concept of the rate of adoption of CAD may be of only limited meaning. Thus, the initial adoption of CAD appeared to be largely influenced by demand-side factors, with both the current level of demand and expectations exerting a positive influence. Even more so than in the case of CNC, adopters of CAD tended to consider only the technical advantages of the technology, accepting that a formal investment appraisal was virtually impossible.

Finally, the results in this paper could be used to provide a broad interpretation of the influence of government policy in the area of new technology. Since 1979, government industrial policy in the U.K. could be interpreted as being very much supply-side in emphasis. However, the conclusions that could be drawn from the results presented here is that, at least in the mechanical engineering industry, demand-side influences would appear to be the stronger. Thus, reliance on supply-side policies to stimulate the adoption of advanced manufacturing technology may have been somewhat misplaced, particularly so to the extent to which these policies were pursued in isolation of the role played by demand. Perhaps the most disturbing policy implication derives from the influence that demand has on the **rate** of adoption of advanced manufacturing technology. Thus, while the supply-side plays some role in the initial adoption decision, demand factors dominate the rate at which adoption takes place. Therefore, to the extent to which an inadequate rate of adoption of advanced manufacturing technology can explain the industry's declining international competitiveness, continuing demand pressures will only serve to exacerbate such forces.

5. REFERENCES

Chow, G. C. (1967). "Technological Change and the Demand for Computers," **American Economic Review, 57**, 1117-1130.

Croxton, E. C., D.J. Cowden and S. Klein. (1968). **Applied General Statistics**, Third Edition. London: Pitman.

Davies, S. (1979). **The Diffusion of Process Innovations.** Cambridge: Cambridge University Press.

Lintner, V. G., M. J. Pokorny, M. M. Woods and M. R. Blinkhorn. (1987). "Trade Unions and Technological Change in the U.K. Mechanical Engineering Industry," **British Journal of Industrial Relations, 25** (1), 19-29.

Mansfield, E. (1968). **Industrial Research and Technological Innovation.** New York: Norton.

Metcalfe, J.S. (1981). "Impulse Diffusion and the Study of

Technical Change," **Futures**, **5**, 347-359.

Nickell, S. (1978). **The Investment Decisions of Firms.** Cambridge: Cambridge University Press.

Pavitt, K., M. Robson and J. Townsend. (1987). "The Size Distribution of Innovating Firms in the U.K.: 1945-1983," **Journal of Industrial Economics**, **35** (3), 297-316.

Pindyck, R. S. and D. L. Rubinfeld. (1981). **Econometric Models and Economic Forecasts**, Second Edition. New York: McGraw-Hill.

Pokorny, M. (1987). **An Introduction to Econometrics.** Oxford: Basil Blackwell.

Ray, G. F. (1984). **The Diffusion of Mature Technologies.** Cambridge: Cambridge University Press.

Romeo, A. A. (1975). "Interindustry and Interfirm Differences in the Rate of Diffusion of an Innovation," **Review of Economics and Statistics**, **57**, 311-319.

Stoneman, P. (1981). "Intra-Firm Diffusion, Bayesian Learning, and Profitability," **Economic Journal**, **91**, 375-388.

Stoneman, P. (1983). **The Economic Analysis of Technological Change.** Oxford: Oxford University Press.

Stoneman, P. and N. J. Ireland. (1983). "The Role of Supply Factors in the Diffusion of New Process Technology," **Economic Journal (Conference Papers Supplement)**, **93**, 66-78.

Tompkinson, P. and M. Common. (1983). "Evidence on the Rationality of Expectations in the British Manufacturing Sector," **Applied Economics**, **15**, 425-436.

A SYNTHESIS OF METHODS FOR EVALUATING INTERRELATED INVESTMENT PROJECTS

James M. Reeve
Department of Accounting and Business Law
University of Tennessee
Knoxville, TN 37996

William G. Sullivan
Departments of Industrial Engineering and Operations Research
Virginia Polytechnic Institute and State University
Blacksburg, VA 24061

ABSTRACT

Strategic investments are frequently placed in increments through time, while simultaneously impacting other processes. These sources of dependency cause the economic evaluation of these projects to be a complex affair. There are a number of methods that allow the analyst to evaluate these types of investments. This paper identifies, explains, and provides practical examples for seven different evaluation methods. For each of these approaches we identify strengths, weaknesses, and best application. We conclude by suggesting that an approach that combines simulation with activity-based costing for tactical evaluation and expert systems for strategic evaluation may have the most future promise.

1. INTRODUCTION

The capital investment decision of the world-class manufacturing firm will require investment justification tools different than those used in the past. Most projects being considered can no longer be adequately evaluated on the basis on their stand-alone benefits. The benefit streams from integrated strategic investments are interrelated because of their interaction with other investments within the firm and through their association with other investments across time. In this paper we will review some of the capital investment tools that are emerging for evaluating the project portfolio.

2. CONCEPTUAL MODEL

A project portfolio consists of multiple capital projects whose benefits and risks are dependent, wherein the acceptance of one project directly impacts the probability of accepting other projects. Investment projects that have an impact on other investment projects can be classified by the degree of the association. The possible degrees of dependency between projects can be expressed as a continuum from strict complements to mutually exclusive. Between these two extremes projects may be negatively related (substitutes), not related in any way (independent), or positively related (complements) as shown in Figure 1. Both perfect complements and mutually exclusive projects are dependent at a point in time. In either case the acceptance of one project has an immediate impact on another. Examples of these dependencies are provided in Table 1.

Figure 1. Continuum of Degrees of Dependence Between Pairs of Projects

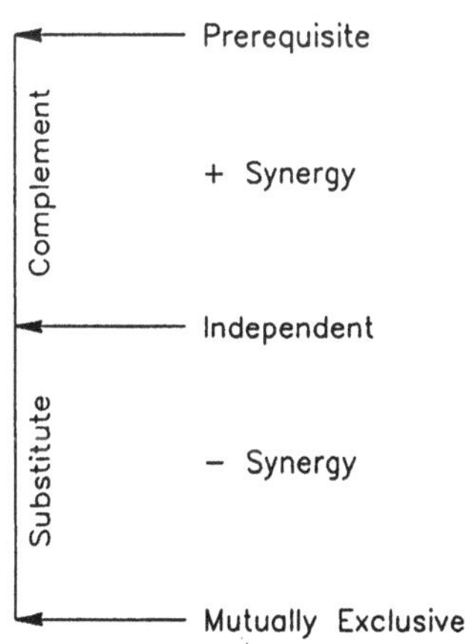

In addition to the nature of the dependency noted in Figure 1, projects can also be categorized in terms of the time phase of the project introductions. Projects can be: (1) dependent across

Table 1. Examples of Dependence Between Pairs of Projects

If the results of second project would ___________ by acceptance of the first project...	then the second project is said to be _________ the first project.	Example
be technically possible or would result in benefits only	a <u>prerequisite</u> to	DNC technology purchase only feasible with previous purchase NC machine tools.
have increased benefits	a <u>complement</u> of	Additional hauling trucks more beneficial if automatic loader purchased first.
not be affected	<u>independent</u> of	A new engine lathe and fence around the warehouse.
have decreased benefits	a <u>substitute</u> for	A screw machine that that would do part of the work of a new lathe.
be impossible	<u>mutually exclusive</u>	A proprietary broad-band cable system for factory communication or an an OSI-based system.

a span of time (cross-temporal investments), or (2) independent of time. Figure 2 is a three dimensional schematic illustrating the conceptual structure of project portfolio analysis. Investment performance on the vertical axis is plotted against operations/processes required to manufacture product on the horizontal axis and the time element on the third axis. Performance is defined as the percentage accomplishment of strategic and/or tactical objectives of the firm after they have

Figure 2. Conceptual Representaion of Primary Issues

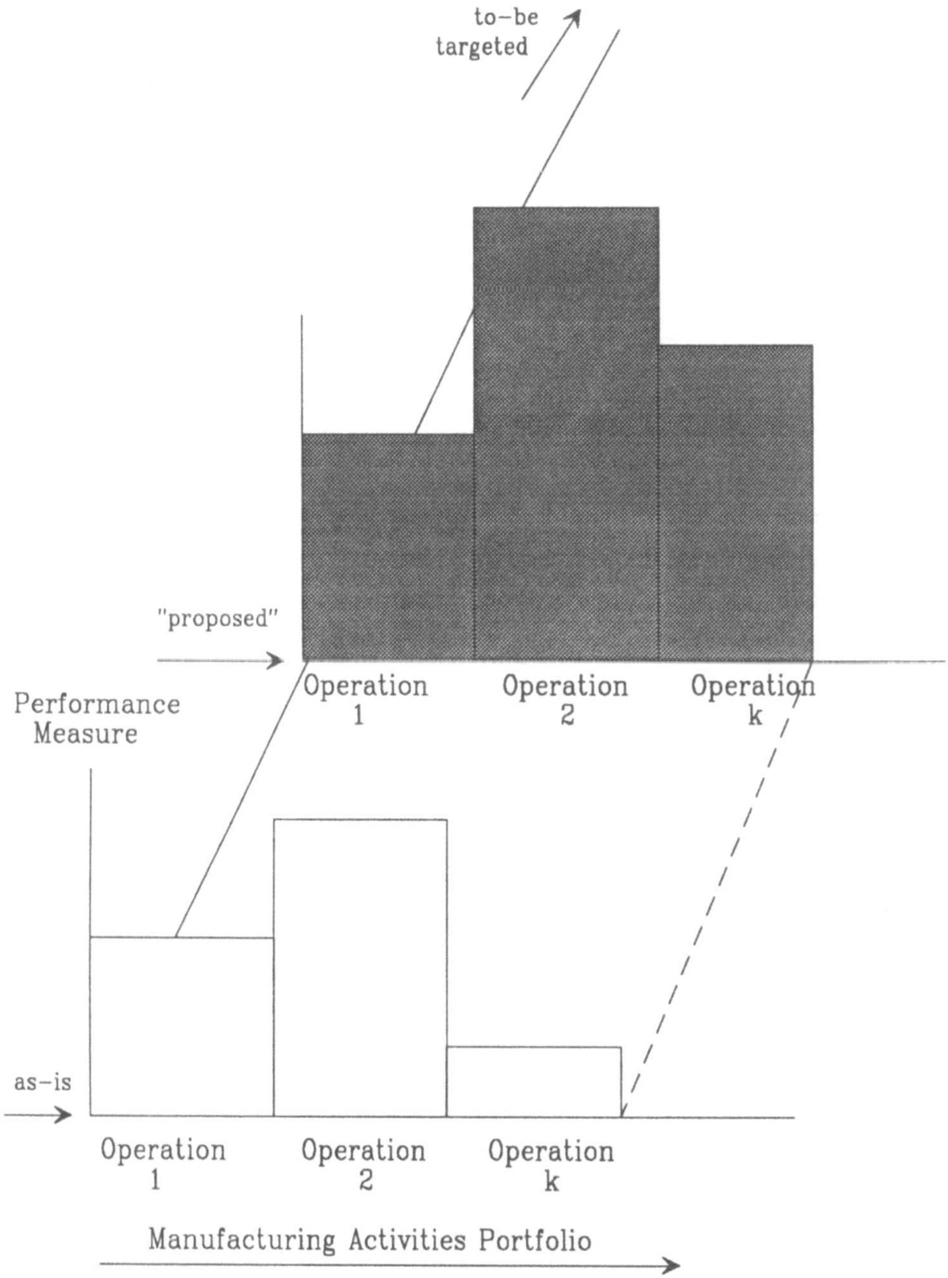

been translated into measurable quantities. Operations are related to each other both at a point in time (cross-sectional) and across time (cross-temporal). The objective is to maximize the expected return and minimize risk of a project bundle along both these dimensions. The target objectives can be obtained through selecting the optimal investment path through time. Clearly, this is a non-trivial analytical problem.

3. TECHNIQUES APPLICABLE TO PORTFOLIO ANALYSIS

The investment evaluation techniques described below all attempt to formally incorporate either cross-sectional or cross-temporal associations (or both) with varying degrees of precision. The appropriateness of each of the approaches are dependent upon the nature of the investment problem. For example, the investment approach is strongly influenced by the degree to which the impact of the investment is understood by the analyst. If the impact of an investment is well understood then the optimal path can be analytically described through constraints, objective functions, and decision trees. If not, then simulation approaches may substitute for explicit formalization of the problem. There is not a one best approach, but rather an array of evaluation choices that are best suited for particular scenarios. The strengths and weaknesses of each approach are identified in our discussion below. In the conclusions we provide the reader with our opinion concerning the investment justification methodology that appears to have the most promise.

4. MATHEMATICAL PROGRAMMING

4.1 Integer Programming

The mathematical programming solutions for interrelated projects generally attempt to discover the bundle of projects which maximize an objective function subject to a number of constraints (see Weingartner, 1966). Integer programming uses integer constraints for complementary, mutually exclusive, and conditionally dependent projects, and assumes that the project cash flows are deterministic (non-random). Because of this assumption, the integer solutions fails to capture the riskiness of project bundles, where risk is measured by the variance of the cash flow return stream from the portfolio.

For example, consider a firm that has available three basic strategies, denoted X_a, X_b, X_c, involving the linking of processing centers. The initial state requires manual movement at two locations, T1 and T2. The solutions may be to incrementally improve by automating T1 and T2 sequentially or by making a single investment in a conveyor that replaces T1 and T2 simultaneously. The mathematical program to solve this problem may be as follows:

$$\text{Max.} \quad \Sigma_j \Sigma_t \, \Gamma_{tj} (1+i)^{-t} \, X_j$$

$$\Sigma_j \, C_{tj} X_j \leq B_t \qquad \text{(budget constraint)}$$

$$X_b - X_a \leq 0 \qquad \text{(prerequisite project)}$$

$$X_a + X_c \leq 1 \qquad \text{(mutually exclusive projects)}$$

$$X_a,\ X_b,\ X_c = 0,\ 1,$$

where Γ_{tj} are the cash flows of alternative j in time t, C_{tj} is the cost of project alternative j in time t, B_t is the budget constraint in time t, and X_j is the integer funding variable for project j (1 accepted, 0 rejected).

The integer program captures the interrelationship between the investment alternatives embodying the two strategies. Namely, investments X_a and X_c cannot occur simultaneously, they are mutually exclusive. The integer constraint guarantees that either X_a, X_c, or neither is accepted; but never X_a and X_c together. Likewise the program also requires that the investment in X_b follows the investment in X_a. X_a is a strict prerequisite to X_b.

4.2 Dynamic Programming of Risky Interrelated Projects

There has been considerable effort to improve upon the deterministic model by incorporating the assumption that the cash flows are random. This assumption allows the risk of the project to be evaluated explicitly. The dependency between two or more cash flows has a non-additive effect upon the risk of the project bundle. The variance (risk) of the cash flows are the sum of the individual project cash flow variances plus two times the covariance of all project combinations. The nature of the covariance term is driven strongly by the correlation between the two project combinations. The net effect is that positively correlated cash flows between individual projects in the portfolio will increase the risk of the portfolio, the opposite would be the case of negatively correlated cash flows.

Hillier (1969) provides an extensive discussion of a dynamic programming solution when the cash flows are considered random. The procedure requires the analyst to combine projects into mutually exclusive portfolios. For each portfolio the mean and variance of the Net Present Values (NPV's) are determined. The variance of the NPV is adjusted for the effect of cross-correlated cash flows. For each project bundle, the mean and variance is determined. The optimal portfolio results when utility is maximized subject to feasibility constraints.

An example of a mathematical programming solution to a complex capital budgeting decision is reported by Park and Son (1988). The problem addressed was identifying the relative merits of a cellular layout (utilizing CNC technology) in the manufacture of gears and couplings relative to a batch shop layout with NC machine tools. The mathematical program used unit cost data on a wide variety of production parameters such as setup, material, machine (idle and running), tooling, and software in building the objective function. The benefits of linked systems through the cellular layout was demonstrated in terms of lower inventory, lower scrap, lower set up times, and improved capital utilization in their example.

4.3 Advantages of Mathematical Programming

1. Requires the analyst to explicitly state assumptions, nature of relationships, unit variable costs, unit revenues, and budget and capacity constraints. This introduces analytic discipline into the process.
2. The solution is correct to the extent the input into the model is valid.
3. Some programming approaches can be designed to incorporate the risk diversifying aspect of project portfolios.
4. The raw inputs into the model are cash flow derived numbers evaluated under the familiar NPV framework.

4.4 Disadvantages of Mathematical Programming

1. The methods require a complete enumeration of the projects, their interdependencies, and their associated cash flow consequences. This is required for those projects considered presently and in the future. The analyst is required to make objective that which is by its nature very subjective. This clearly imposes severe constraints on the applicability of these approaches in practice.
2. The model outputs are highly conditioned on the reliability of the data input. As a result management must have a detailed understanding of their present and future options before these techniques can be applied. For example, in dynamic programming an explicit articulation of all possible outcomes are required in order for the method to optimize a decision path. The potential combinations in a real world setting can quickly become burdensome.
3. Some of the formulations ignore risk.
4. Programming solutions generally accommodate only one objective function, to maximize NPV. There may be other strategic objectives that require enumeration beyond the financial returns.
5. For large combinations of projects the programming approach quickly becomes unwieldy.
6. Many of the models suffer from the common deficiency of assuming that the reinvestment rate will be the discount rate.

5. OPTION VALUE ANALYSIS

One of the limitations of traditional investment justification approaches is the failure to consider the value implicit in the active management of projects. By active management we mean the prerogative of management to change the course of action of an existing project by either abandoning the project, lengthening or shortening the investment commitment time period, coupling existing projects with new projects to achieve future growth, committing investments to maintain flexibility, or any other type of contingent-based planning that represent options to the firm. These contingencies are not included in traditional evaluation criteria and so the value of most projects is generally understated. New tools in the area of financial option valuation provide potential for explicitly incorporating these options into the justification analysis.

Consider a firm that invests in an R&D project or technological improvement for their production process. A question arises: do these investments provide opportunity for additional returns due to their synergistic rollover effects on other opportunities? The answer must consider how a firm creates value. Long-term value may be created through a strategic process of linking connected time phased investments. At the time of the initial investment the opportunities for synergy are unknown, but are assumed to exist. In other words, the investment provides value on its own merits and, in addition, through the flexibility it generates for future return options. These options are conditional upon making the initial investment commitment and can create greater value by providing even further options (compound options). Benefits are realized through opportunities for future product or process learning, growth, product awareness, asset replacements/improvements, market expansion, and the like (Kester, 1984).

The option on future opportunities may be modeled as a financial call option. A call option is the right (but not the obligation) to purchase the benefits of an asset at a specified price (exercise price) within a certain frame of time. After the time period has expired the option cannot be exercised and becomes worthless. Any opportunity for strategic options is conditional on making the first stage investment. This option value may be totally firm specific, and so it need not be reflected in the purchase price of the investment. The value of the option on future opportunities depends upon the value of those opportunities today (the existing market price) and the exercise price (the investment price).

This area of financial analysis is still very young and has only been applied to more simplified settings to date. For example, the abandonment option has been characterized as a put option (McDonald and Siegel, 1985, 1986). The investment carries the option to sell the project whenever the project goes sour. Therefore, the option creates additional value by limiting downside risk.

Another example is the use of option pricing theory to evaluate the decision to continue running a gold mine. The decision is viewed as an option to buy gold with the exercise price representing the cost of running the mine. Brennan and Schwartz (1985) successfully applied this technique in such a setting. Still another example is the use of option pricing theory to determine the option value of being able to delay long term irreversible sequential investment expenditures, such as building a plant (Majd and Pindyck, 1987).

5.1 Advantages of Option Value Analysis

1. Has very strong theoretical support as an alternative to cash flow forecasting. The method uses existing market values in place of discounted NPV as the basis of analysis. Removes a great deal of subjectivity.

2. Captures the value of time phased active management in the project selection decision.

5.2 Disadvantages of Option Analysis

1. The methodology is being applied in only limited situations where market values are obtainable relative to the investment decision. Without known market values the method is not presently applicable. This is a severe restriction for many

capital budgeting problems.

2. The method is analytically unfamiliar to many and therefore may be subject to "black box" misuse.

6. SCORING MODELS

A popular technique for subjectively evaluating multiple objectives is a scoring model approach. Scoring models are analytical approaches that weight the subjective criteria of an investment decision. For example, scoring models allow the analyst to subjectively incorporate the impact of quality, flexibility, lead time, reliability, schedule stability, and risk on the investment portfolio. The weighting scheme for each subjective attribute can be unweighted, meaning the attributes being subjectively evaluated are considered equal in importance or weighted, which allows the attributes be have different importance factors.

An example of a weighted scoring model approach is the Multi-Attribute Decision Model (MADM). MADM has been applied by McGinnis, et al. (1985) to evaluate alternative tooling configurations for a FMS under conditions of demand uncertainty. Their MADM approach specifically incorporated subjective assessments of customization and size flexibility, throughput, and repair statistics. Another application of the MADM scoring model approach is demonstrated by Sullivan and Liggett (1988). In their approach a software tool (called JUSTLAN) was developed to aid in the justification of a LAN. The software elicits decision maker scoring evaluations for project attributes such as security, flexibility, expendability, and the like. The weighted importance scores are used to develop an overall performance score for each investment alternative. These scores are combined with financial considerations to provide the analyst with a graphical interpretation of the relative merits of two competing LAN investments on both financial and qualitative dimensions.

Analytical Hierarchy Process (AHP) is another scoring method developed by Saaty (1980) to capture the intuitive judgements of decision makers for decisions involving complex interactions. The methodology uses pairwise comparisons to elicit judgements about the relative importance of project characteristics within a stated hierarchy. The result is an overall set of priorities that can be used to select project alternatives. Varney et al. (1985) illustrate the application of the analytical hierarchy process (AHP) to the evaluation of a flexible manufacturing system. The method was used to evaluate pairs of qualitative characteristics within a hierarchy of benefits and costs that included global, strategic, and operational considerations.

6.1 Advantages of Scoring Models

1. Easy to understand.
2. Incorporates financial and non-financial measures (i.e. qualitative characteristics are explicitly considered).
3. Attempts to coordinate operational plans with strategic objectives.
4. Requires a subjective rating of operational concerns and degree to which investment alternatives satisfy those operational concerns.
5. Risk is not ignored, but is subjectively evaluated.
6. Accommodates multiple objectives.

6.2 Disadvantages of Scoring Models

1. All inputs into the model are subjective.
2. The outputs of the scoring model are not subject to rigorous defense.
3. Explicit analysis of interrelated projects and their impact on flexibility and risk are ignored.
4. Disconnects the present value criterion from subjective assessments that impact present value, but are difficult to quantify.
5. Assumes independence of the operational factors.
6. Output scores of the model can only be interpreted as relative measures. The scores have no absolute meaning in themselves.

7. SIMULATION

7.1 Factory Simulations

Simulation can be used to evaluate and help eliminate the risks involved in capital budgeting. Initially, several projects might be considered for a production environment. Before costs are considered, each option can be evaluated through a simulation of the production process. The basic input structure of a production sequence can be modeled through simulation software, from which output characteristics such as lead time, throughput, average inventory, and workloads can be determined. Moreover, simulation enables exploration of output sensitivity to changes in such underlying input variables as operating times, operating time variability, scrap rates, equipment downtime profiles, setup times, and scheduling loads/variation (Vester and Muller, 1987).

A simulation of complex interrelated manufacturing improvements such as new processes, facilities, and layouts can help identify and eliminate mutually exclusive alternatives that have unapparent deficiencies. One of the advantages of simulations is that the results can be determined without completely identifying the mathematical relationships and correlations that exist within the proposed investment bundle.

There are many examples in the literature of the use of simulation software in justifying complex interrelated investments. Warnecke and Vettin (1982) simulate alternative workpiece flows in a FMS using a program called MUSIK. The simulation identified the optimal batch size, workpiece carriers, and Work in Process (WIP) storage capacities in order to have optimal unit costs under various mix and demand conditions. Suresh and Meredith (1985) used SIMSCRIPT to simulate a multi-machine FMS. Their simulation identified economic drivers relating capacity requirements to part volume/variety.

Hutchinson and Holland (1982) use simulation to demonstrate the value of adding capacity in small increments. Value results from avoiding abandonment of expensive investments when demand or product characteristics change through their life cycles. This is a similar objective to the option pricing approach discussed previously. In another study, Sullivan et al. (1989) evaluate a complex investment strategy in the continuous process industry. Their simulation identified the specific tradeoffs between running geographically disperse plants with focused vs. mixed product strategies.

Factory simulations were confined to fairly arcane programming languages such as GPSS that require specialized programming skill. More recently, user friendly factory modeling programs are becoming available to the analyst. Such programs as XCELL+ (Conway, et al., 1988) use embedded menus so that the factory can be "built" on the screen using various predefined factory elements (workcenters, buffers, maintenance facilities, and so forth). An excellent example of the use of XCELL for simulating various factory scenarios is provided by Monahan and Smunt (1987). Such tools do not require sophisticated programming skills, and can lead to "rough cut" analysis of the factory in a fraction of the programming time that was needed previously. These new tools will undoubtedly make simulation a popular method to evaluate the complex interactions inherent in process changes.

7.2 Monte Carlo Simulation

The factory simulation provides operating information that can be used as input to a Monte Carlo simulation. Monte Carlo methods are used to evaluate the impact of change in environmental factors that may impact the investment. Such factors would include demand, sales price, operating costs, life of project, and selling costs.

The Monte Carlo simulation procedure in capital budgeting involves the following basic steps as described in detail by Hertz (1964).

1. For each input factor estimate the range of values and their likelihood for occurrence from subjective assessments or production simulation data.
2. Select at random one value from the distribution of values for each factor.
3. Combine the values for all input factors and compute the

associated cash flows and NPV.

4. Repeat this process over and over by sampling from the value distributions of the input factors.

The repetition inherent in step four results in a cumulative probability density function of NPV's. Alternative projects can be selected based upon their stochastic dominance. Namely, if the cumulative probabilities of achieving all levels of NPV for project one is greater than project two, then it is said that project one dominates project two. This concept is illustrated by Meredith and Suresh (1986) using a simplified example. Practical application of the Monte Carlo approach has been documented by Kryzanowski, et al. (1972) for incremental plant investments over time with dependencies among factors within a period of time.

7.3 Advantages of Simulation

1. The simulation can be built from known data without understanding all aspects of the problem. For example, simulation approaches do not require that analytical correlation information between time-phased investments and interaction effects on the production floor be specified completely.

2. The method is can be easy to use with popular software now available.

3. The output is easily interpreted since it is expressed in operational terms, such as lead times and utilizations.

4. The analyst can explore changes in the underlying variables by using sensitivity analysis on such things as production times, cycle times and scrap rates.

5. Simulation formally incorporates risk, and provides an array of possible outputs over many runs. Therefore, simulation allows the analyst to determine the expected value and variance (risk) of returns.

6. Production simulations are very good at evaluating complex interactions from investments on the floor. Monte Carlo simulations can be designed to evaluate time-phased investment scenarios.

7.4 Disadvantages of Simulation

1. The values for the simulation factors and their probability distributions are subjectively estimated for Monte Carlo simulations. The same is also true for some of the production simulation parameters. Investment justification is only possible if the analyst is informed about the impact of the investment on the environment.

2. Both simulation approaches require some specialized skill in formulating the investment problem in a form that can be computer generated.

3. The approach is a customized, rather than a generalized approach. That is, a different simulation is required for each specific investment situation. In this sense, the approach is expensive to use.

8. DECISION FLOW NETWORKS

Another useful method for evaluating interrelated projects, that are encountered as sequential multi-stage problems involving a series of time-phased decisions, is decision flow networks or "decision trees" (Canada, 1974 and Magee, 1964). Figure 3 illustrates the basic characteristics of the decision flow network, including decision nodes and chance nodes. Emanating from each decision node is a set of branches, each representing one of the strategies available for selection. The decision nodes lead to chance nodes along these strategy branches. Each chance node is followed by a set of outcome branches and associated probabilities. From these outcome branches the expected value and risk of the strategy can be determined. In Figure 3, alternative A is a strict prerequisite to alternative B for strategy 1. For strategy 2, alternative B is obtained immediately without prior sequencing of alternative A. Note that the higher expected value of the immediate move to B comes at greater risk.

The decision flow network is particularly adept at modeling investments over time. Time is modeled as a series of sequential decision branches from left to right in Figure 3. Each time-phased decision has outcome probabilities which then lead to the next period's decision branch. The number of "real world" outcomes emanating from an alterative is greater than the few outcomes

Figure 3. A Decision Flow Network

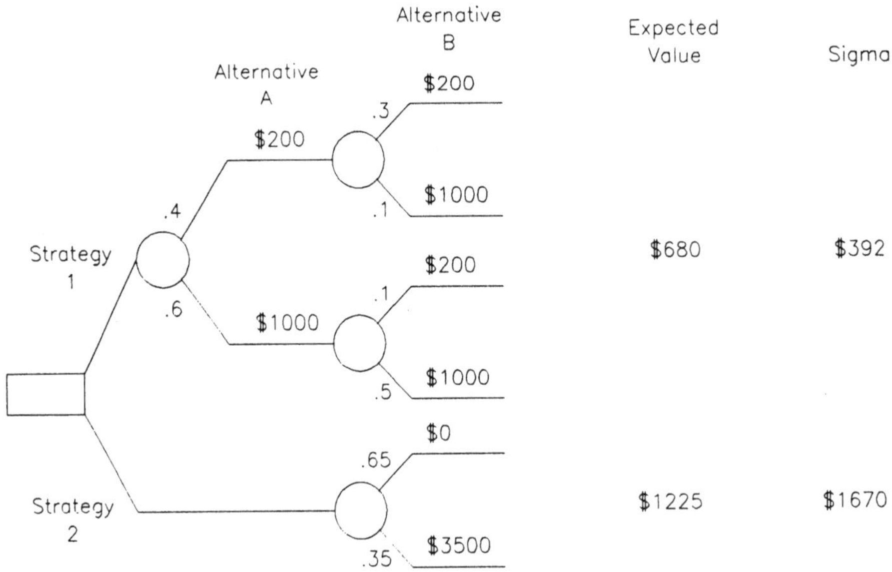

typically assumed in order to manage the size of the tree. Information, and sometimes, objectivity, may be sacrificed by the assumed reduction in decision branches. Often the outcomes emanating from a strategy branch can be best described by a continuous distribution. Such networks can readily be analyzed using Monte Carlo simulation techniques as described earlier.

The decision flow network incorporates more data then the integer program. For example, the integer program did not take into account the uncertainty in cash flows, and so did not provide a mechanism for measuring the relative risk of the various strategies. Unfortunately, one of the disadvantages of the decision flow network is the sheer size and complexity real problems impose on network design.

8.1 Advantages of Decision Flow Networks

1. Makes uncertainty explicit through specifying the outcome branch probabilities.

2. Encourages consideration of the complete problem through time, and provides the optimal investment path (similar to dynamic programming).

3. Provides information about the riskiness of the project.

4. Stimulates discussion and consideration of new alternatives and possible outcomes.

8.2 Disadvantages of Decision Flow Networks

1. Requires a high degree of data specification, which can be beyond the knowledge of the analyst. Often the analyst would rather take refuge in the fuzziness and complexity of real-life situations than to reveal preferences in a starkly-simple decision structure.

2. Requires expert articulation of the thought process, and so may restrict the complex synthesizing of interconnected considerations that occur subconsciously.

3. Tends to exclude the consideration of intangibles through the formalization of decision rules.

4. Is very unwieldy in large complex problems involving many decision and chance nodes.

9. ACTIVITY BASED COSTING

One of the newer innovations in accounting is the rise of activity based costing (Cooper and Kaplan, 1988 and Turney, 1989). Activity based costing (ABC) replaces traditional accounting methodologies by determining the critical cost drivers of an organization and relating these drivers to both process and product. Traditional cost accounting usually assumes very naive relationships between product and process, such as allocations based on machine hours or direct labor hours. Whereas, ABC determines the critical relationships between the activities of the firm and product manufacture. ABC systems use cost drivers based on such activities as the number of setups, number of orders, number of inspections, number of part numbers in the product design, number of production runs, and number of material receipts. These cost drivers represent a cost rate for performing various activities. For example, a setup triggers cost. Under ABC a setup occurs at a particular cost rate per setup activity. Product would be charged this rate for every lot setup.

The activity based cost system architecture should be an excellent tool for investment justification (see Reeve, 1989 and Blank, 1984). Investments change the process, and therefore change the activity structure. The impact on cost can be assessed by modeling the new activity structure at the present activity rates. For example, consider an investment in a focused factory layout. This layout should sharply curtail the number of expedited items. Specifically, the level of this activity will be reduced. If the present investment level is unchanged then the activity will become over-resourced. Expediting will have a reduced denominator level of activity, so the dollars invested in performing expediting will have to be reduced, or else there will be a large capacity cost in this activity.

The ABC system gives the engineer the ability to evaluate process changes in terms of the impact on resourcing activities of the enterprise. Over-resourced activities that emerge from investments in new processes become visible areas for cost reduction and are the source of cost savings in the investment justification. Simulation tools could be very helpful in identifying the activity level changes that would occur from investments.

The ABC will indicate that some investments will need further process changes in order to accommodate new activity levels in other parts of the plant. For example, an investment in setup improvements will allow the manufacturer to process in smaller lots. However, the cross-activity impact of improved setups will be to increase the number of raw material receipts, number of inspections, number of shipments, and number of handling transactions due to the smaller lot sizes. If the technology of moving material, inspecting, receiving, or shipping does not change, these activities will become under-resourced. The present investment levels in these activities will be insufficient to support the new increased physical demands. The ABC clearly begins to guide the firm into investment portfolios that improve many processes simultaneously. Using our example, the full benefit of the improved setup must be coupled with process changes that improve lot-related transactions (e.g. on-line SPC, EDI receiving, small distance layouts). The ABC can now estimate the full cost savings from reducing the resourcing levels of all these critical activities once the process changes are in place.

9.1 Advantages of Activity Based Costing

1. Specifically associates cost to improvements in the process. This approach completes the link to production simulation, which provides operational data from process changes. ABC determines the cost impact of these operational changes.

2. Provides the analyst with insight to where other complementary investments may be needed to support a given process change.

3. Provides information on both over- and under-resourced activities.

9.2 Disadvantages of Activity Based Costing

1. Activity based costing systems are fairly new, and therefore needed software is generally unavailable.

2. The ABC has fairly extensive data requirements that may add some information complexity to the organization.

3. The use of ABC in investment justification has not been widely demonstrated; therefore, both weaknesses and strengths are unproven.

10. ARTIFICIAL INTELLIGENCE AND KNOWLEDGE BASED EXPERT SYSTEMS

One of the newer areas of interest is using computers to mirror decision rules of human problem solving. Knowledge Based Expert Systems (KBES) have been the success story of artificial intelligence because they allow a more flexible approach to problem solving. Many real life problems cannot be represented accurately by an analytical model or algorithm. Some problems are better modelled by heuristic techniques, which are the formulating strategies of experts based upon a structured background of expertise. The use of expert systems for investment portfolio analysis has been suggested and illustrated as a viable technique (Sullivan and Reeve, 1988).

The development of KBES for investment situations provides the advantage of simulating the human problem solving and judgement expertise of an expert. The knowledge base can be formulated by a series of IF-THEN rules that can be interactively exchanged between the user and the system. The user need not be an expert in the decision setting, but merely sufficiently informed to respond to the queries of the system. Naturally, the approach assumes that experts in the appropriate decision domain exist and that their decision policy rules can be captured by the expert system architect.

There have been many applications of the use of expert systems in the manufacturing environment. Most of these applications involve the use of expert systems as diagnostic tools for complex manufacturing processes. For example, an expert system could interface with a production control system and provide expert guidance on trouble shooting and process control. The use of expert systems in the financial arena has been well documented (Clifford et.al., 1986 and Brown, 1988), but few have been developed for the investment justification decision. Sullivan and Reeve (1988) illustrate the use of XVENTURE as a expert system advisor for the justification decision. The system is based on the reasoning of a single expert along six strategic considerations. After the user responds to a series of queries, XVENTURE then provides a go, no-go, or defer decision based upon expert decision rules combining the six dimensions. Dilts and Turowski (1988) develop a strategic advisor for investment justification using a Porter (1980) analysis framework. This system evaluates the impact of the investment relative to sustaining competitive technological advantage. Again, the expert advisor is built around a series of queries about features of the sustainable technological advantage. The user responses to the queries are applied to heuristic rules built into the computer software from which a recommendation is then provided.

The KBES effort to-date has been very encouraging. Although there is little experience in the investment justification decision domain, one insight is emerging. The use of KBES will probably be limited to evaluating strategic considerations rather than the technical considerations of the project (Demetrius, 1986). The nature of expert rules for evaluating technical considerations are likely to be so project specific as to make the development of a KBES impractical. In contrast, the broad strategic considerations that are necessary precursors to the technical evaluation are much more adaptable to policy-capturing approaches. As a result, KBES's will probably emerge as a front-end of a technical evaluation of the project portfolio, with decision trees or simulations used in a second stage. The KBES would provide widely available expertise throughout the firm for purposes of evaluating the strategic considerations of technology proposals.

10.1 Advantages of KBES's

1. Allows expertise to become widely available in the organization.

2. Formulating the KBES is an instructive exercise in forming rules that incorporate strategic considerations. The expert him(her)self even becomes more aware of the heuristics used in strategic considerations. Likewise, the KBES becomes a method of infusing strategic analysis and awareness into the organization.

3. Causes an explicit consideration of strategic consequences apart from the discounted cash flow model. This is a powerful advantage because some of the strategic issues are difficult to capture in traditional project evaluation methodologies.

10.2 Disadvantages of KBES's

1. KBES's will probably only be appropriate for use in broad strategic evaluations, not technical evaluations.

2. KBES's are time consuming and expensive to develop. They are subject to obsolescence, and so must be maintained and updated to incorporate new strategic considerations.

3. A KBES may cause the firm to become fixated on the present expert decision rules. The KBES rule structure may lag environmental changes and cause poor selection decisions.

4. There is some concern that expertise is a function of

experience. If followed too rigidly, KBES's minimize the value of experience. As a result, the use of KBES's may retard development of the next generation of experts.

11. CONCLUSIONS

Table 2 provides an appraisal summary of the approaches that have been reviewed and illustrated. The analysis of interdependent projects is by its very nature a difficult analytical issue. As such, some of the methods are better than others in accommodating this complexity. At one end of the spectrum are the mathematical programming techniques. These approaches require that the relationship between projects be explicitly enumerated. The solution techniques are sound if indeed the inputs can be determined. The difficulty, of course, lies in delineating the nature of the dependencies. The risk of the portfolio cannot be determined explicitly without an estimate of the covariance structure of returns in the individual project groupings. This in turn requires an estimate of the correlations between project cash flows, a challenging metric to determine, especially if the number of interdependencies grows much beyond some small number.

Close to mathematical programming approaches in terms of the input demands are the stochastic decision flow network analyses and simulation approaches. Both of these also require a high degree of input specificity. They have an advantage over the mathematical programming techniques, however, in that the output is visual and interpretable. Furthermore, the methods are amenable to sensitivity analyses, and hence risk assessment.

The option valuation approach is presently in early development. For those areas where the approach can be used the data inputs are not overly burdensome. We believe, however, that the use of this technique for valuing flexibility through the option of future investment opportunity is a long way off. There is some promise, however, for using this technique for discrete project bundles that convert market-priced inputs into market-priced outputs. The natural resource industry would be an application area with high promise.

We believe the most promising area of development will involve

TABLE 2. Appraisal Summary of Portfolio Analysis Techniques

Portfolio Technique	Principal Advantages	Principal Weaknesses	"Best Fit" Application
A. Math Programming	Consistent with NPV concepts. Familiar to many users. May be able to give objective maximizing investment decisions at low levels of project interaction.	Many formulations are deterministic. Very complex at moderate levels of project interaction. Requires a high degree of input specificity. Controversy over the objective function.	Analysis where there is a low level of project interaction, and correlation between project cash flows can be estimated.
B. Option Valuation	Explicitly incorporates the value of flexibility. Investment may be valued using extant market data and avoid present value assumptions.	Asset prices underlying options are known, therefore method is unable to strictly value options on future investment opportunities.	Abandonment valuation. Natural resource investment analysis. Sequential investment timing in long term projects.
C. Scoring Models	Simple formulation. Accommodates multiple objectives.	Overly simplifies problem. Extremely subjective.	First cut screening to select projects consistent with objectives of corporate strategy.
D. Simulation	Can make changes to the problem easily. Can evaluate interrelated projects and the interrelationships between variables within one project. Does not require probability distributions to fit any classical distribution.	Often difficult to develop an accurate model of the decision situation. Difficult to estimate the data required for input.	Applicable to multiple scenario evaluations under conditions of risk. Interdependencies can be modeled through modification of cash flow estimates.
E. Decision Flow Networks	Makes uncertainty explicit. Promotes more reasoned estimating procedures. Encourages consideration of whole problem and helps communication. Stimulates generation of new alternatives. Provides framework for contingency planning.	Tends to exclude consideration of non-monetary factors. The clear basic questions are often most difficult to articulate.	Large and complex investment problems that are time dependent are ideal for decision flow diagramming. Risk can be introduced by considering chance nodes to be stochastic.
F. Activity Based Costing	Determines cost impact of activity changes.	Moderately complex information support system.	Combine with production simulation output to estimate cost effects.
G. Expert System	Can deal with "fuzzy" information from many experts. Can handle conflicting and interrelated information. Easy to transfer and store knowledge. Consistent and affordable to use.	May give "most probable" or "best" solution but not necessarily the "right" solution. Rule base is inflexible for changing conditions. Choice of an "expert" is critical.	Where the problem requires "heuristic" approach to interrelated projects. Used when facts are known but not precisely; when perishable expertise is expensive but available.

a combination approach. An expert system will help guide the decision maker in evaluating the investment in light of strategic considerations. Production simulations can be used to evaluate the operational characteristics of the investment. The output of the production simulation forms the necessary data for an activity cost analysis. Specifically, changes in the activity structure can be determined from the simulation, while the activity based cost system translates these activity changes into cost determinations. The above suggestion forms three connected levels, a strategic level analysis, an operational analysis, and a cost analysis. Of the three levels, expert systems and activity based cost systems are the least developed. Production simulations are widely used tools. When production simulations are combined with appropriate strategic considerations and activity based cost models, then the analyst will have the ability to make an evaluation across the complete problem spectrum.

12. REFERENCES

Blank, L. (1984). "The Changing Scene of Economic Analysis for the Evaluation of Manufacturing System Design and Operation," **The Engineering Economist, 30** (3), 227-243.

Brennan, M. J. and E. S. Schwartz. (1985). "Evaluating Natural Resource Investments," **Journal of Business, 58**, 135-157.

Brown, C. E. (1988). "Accounting Expert Systems: An Annotated Bibliography," mimeo, Oregon State University.

Canada, J. R. (1974). "Decision Flow Networks," **Industrial Engineering**, June, 30-37.

Clifford, J., M. Jarke, and H. C. Lucas. (1986). "Designing Expert Systems in a Business Environment," **Artificial Intelligence in Economics and Management**. Amsterdam: Elsevier Science Publishers.

Conway, R., W. L. Maxwell, J. O. McClain, and S. C. Worona. (1988). **User's Guide to XCELL+ Factory Modeling System**. Redwood City, CA: The Scientific Press.

Cooper, R. and R. S. Kaplan. (1988). "Measure Costs Right: Make the Right Decisions," **Harvard Business Review**, September-October, 96-105.

Demetrius, D. G. (1986). "Expert Systems and Board Level Decisions," **Artificial Intelligence in Economics and Management.**

Amsterdam: Elsevier Science Publishers.

Dilts, D. M. and D. G. Turowski. (1988). "Incorporating Strategic Knowledge Within Investment Justification," working paper, University of Waterloo.

Hertz, D. (1964). "Risk Analysis in Capital Investments," **Harvard Business Review**, January-February, 95-106.

Hillier, F. S. (1969). **The Evaluation of Risky Interrelated Investments.** Amsterdam:North-Holland.

Hutchinson, G. K. and J. R. Holland. (1982). "The Economic Value of Flexible Automation," **Journal of Manufacturing Systems, 1** (2), 215-228.

Kester, W. C. (1984). "Today's Options for Tomorrow's Growth," **Harvard Business Review**, March-April, 153-160.

Kryzanowski, L., P. Lusztig, and B. Schwab. (1972). "Monte Carlo Simulations of Capital Expenditure Decisions- A Case Study," **The Engineering Economist, 18** (1), 31-48.

Magee, J. F. (1964). "Decision Trees for Decision Making," **Harvard Business Review, 42** (4), 126-138.

Majd, S. and R. S. Pindyck. (1987). "Time to Build, Option Value, and Investment Decisions," **Journal of Financial Economics**, March, 7-27.

McDonald, R. and D. Siegel. (1985). "Investment and the Valuation of Firms When There is an Option to Shut Down," **International Economic Review, 26**, 331-349.

McDonald, R. and D. Siegel. (1986). "The Value of Waiting to Invest," **Quarterly Journal of Economics, 101**, 707-727.

McGinnis, M. S., K. M. Gardiner, and R. Jesse. (1985). "Capital Equipment Selection Strategy Under Volatile Economic Conditions," **AUTOFACT Conference Proceedings**, 67-86.

Meredith, J. R. and N. C. Suresh. (1986). "Justification Techniques for Advanced Manufacturing Technologies," **Justifying New Manufacturing Technology.** Norcross, GA: Institute of Industrial Engineers, 82-98.

Monahan, G. E. and T. L. Smunt. (1987). "A Multilevel Decision Support System for the Financial Justification of Automated Flexible Manufacturing Systems," **Interfaces**, November-December, 29-40.

Park, C. S. and Y. K. Son. (1988). "An Economic Evaluation Model for Advanced Manufacturing Systems," **Engineering Economist, 34** (1), 1-26.

Porter, M. E. (1980). **Competitive Strategy.** New York:The Free Press.

Reeve, J. M. (1989). "Variation and the Cost of Activities for Investment Justification," **Proceedings of the IIE Integrated Systems Conference.**

Saaty, T. L. (1980). **The Analytical Hierarchy Process: Planning, Priority Setting, Resource Allocation.** London: McGraw Hill International.

Sullivan, W. G. and H. R. Liggett. (1988). "A Decision Support System for Evaluating Investments in Manufacturing Local Area Networks," **Manufacturing Review, 1** (3), 151-157.

Sullivan, W. G. and J. M. Reeve. (1988). "XVENTURE: Expert Systems to the Rescue," **Management Accounting**, October, 51-58.

Sullivan, W. G., J. M. Reeve, and R. S. Sawhney. (1989). "Strategy Based Investment Justification for Advanced Manufacturing Technology," **CAM-I Research Report.**

Suresh, N. C. and J. R. Meredith. (1985). "Justifying Multimachine Systems: An Integrated Strategic Approach," **Journal of Manufacturing Systems, 4** (2).

Turney, P. B. B. (1989). "Accounting for Continuous Improvement," **Sloan Management Review**, Winter, 37-47.

Varney, M. S., W. G. Sullivan, and J. K. Cochran. (1985). "Justification of Flexible Manufacturing Systems with the Analytical Hierarchy Process," **IIE Annual Conference Proceedings**, 181-190.

Vester, J. and D. Muller. (1987). "Reduce Manufacturing Risk With Simulation," **Automation**, November.

Warnecke, H. J. and G. Vettin. (1982). "Technical Investment Planning of Flexible Manufacturing Systems: The Application of Practice Oriented Methods," **Journal of Manufacturing Systems, 1** (1), 89-98.

Weingartner, H. M. (1966). "Capital Budgeting of Interrelated Projects: Survey and Synthesis," **Management Science 6** (1), 485-516.

A STRATEGIC EVALUATION METHODOLOGY FOR MANUFACTURING TECHNOLOGIES

Fariborz Y. Partovi
Dept. of Management and Organizational Sciences
College of Business and Administration
Drexel University
Philadelphia, PA 19104

ABSTRACT

The purpose of this paper is three-fold. First, a prescriptive methodology relating manufacturing strategy to choice of technology is presented. Second, an experimental evaluation of the proposed methodology is described. Finally, the managerial implications of the methodology are explored in the context of a real-world case study. The forces that shape the firm's competitiveness, the approach used to operationalize the variables in the real world, and the characteristics of the technologies are discussed. Based on the results of the case study, the technological options available to the firm were prioritized.

1. INTRODUCTION

The enormous growth in manufacturing automation has brought a myriad of automated machines with diverse features. The ability to make rational choices among these advanced technologies seems increasingly necessary in today's competitive business environment. This paper proposes a new method for technology evaluation in manufacturing organizations. The methodology is based on the technological planning process of Kleindorfer and Partovi (1990). This paper contributes to the methodology and implementation procedures for technology selection in a manufacturing environment.

The paper first provides a brief outline of prior research on equipment selection and justification. Section 3 discusses the proposed methodology. Section 4 presents the computerized hierarchical model used in this study. Section 5 describes an experiment on how subjects perceive the method and how experts compare the effectiveness of the proposed model with traditional justification methods. Section 6 discusses the implementation of the proposed model in a manufacturing environment. Finally, section 7 concludes the paper with a summary and suggestions for future research.

2. PREVIOUS RESEARCH

Technology choice is fundamental to the planning process in a manufacturing setting. Traditionally, the ranking and choice of projects for technological modernization in a manufacturing environment relied on capital budgeting techniques such as payback, internal rate of return and present value (Diaz, 1986; Swindle, 1985; and Noble, 1989). These techniques have been extensively used in industry because of their ease of application; rational, tactical financial assumptions; and treatment of the time value of money. However, although these models may be sound financial techniques, critics have repeatedly charged that they are limited since: (1) they are "tactical" rather than "strategic" (Park and Young, 1989); and (2) they do not include intangible benefits such as improved quality, increased flexibility and decreased delivery time, each of which increases the revenue portion of the profit equation. Some authors have gone even further to claim that capital budgeting techniques are quite likely the greatest barrier to the implementation of new technologies by manufacturing industries (Kaplan, 1986, and Skinner, 1985).

In recent years the strategic aspects of the choice of technology have been receiving more attention (Jelinek and Golhar, 1983; Blois, 1986; Swamidass, 1987; Kleindorfer and Partovi, 1990; and Stevens and Martin, 1989). Competitive pressures have forced manufacturing managers to recognize the importance of technology in enhancing the quality, flexibility and delivery of products. Most research in this area has been descriptive in nature (Blois, 1986, Jelinek and Golhar, 1983). However, some authors have recently proposed prescriptive models related to choice of technology. For example, Swamidass (1987) proposes a set of indices for cost, quality, and flexibility which measures the deterioration of installed technologies in terms of these forces, and compares them with state-of-the-art technology or with the technology used by the chief competitors. Park and Son (1989) suggest a multistage investment-decision model that considers the nonconventional costs of quality, flexibility and delivery. However, such prescriptive approaches do not integrate the forces driving competition and ignore the changes new production technologies will have on revenues. This paper proposes a new methodology for technology evaluation for manufacturing organizations. The prescriptive methodology proposed begins with the competitive strategy that provides the backdrop for the development of a manufacturing strategy. In particular, the relative importance of cost, quality, flexibility, and delivery for long-run profitability and competitive viability are evaluated. These considerations lead to the development of a performance hierarchy for a given line of business. It is this performance hierarchy which, together with an evaluation methodology based on Saaty's (1982) Analytical Hierarchy Process (AHP), leads to the proposed procedure for prioritizing alternative technologies.

3. TECHNOLOGY EVALUATION METHODOLOGY

The technology evaluation methodology (TEM) suggested in this paper is based on the TPP or Technology Planning Process (Kleindorfer and Partovi, 1990) which is loosely related to Ackoff's interactive planning (Ackoff, 1981). The TPP consists of five stages: external and internal analysis, ends planning, means planning, resource planning, and implementation and control. In Figure 1 we only show the first three stages of this methodology which are relevant to TEM.

The **External analysis** addresses the threats and opportunities facing the firm. In this stage all the stakeholders that are affected or affect the competitiveness of the industry will be identified and their relation to the particular firm of interest will be understood (Porter, 1980, and Freeman, 1983). The **internal analysis** identifies the strengths and weaknesses related to the internal operations and resources of the firm in terms of the value

Figure 1. Methodology of Technological Evaluation

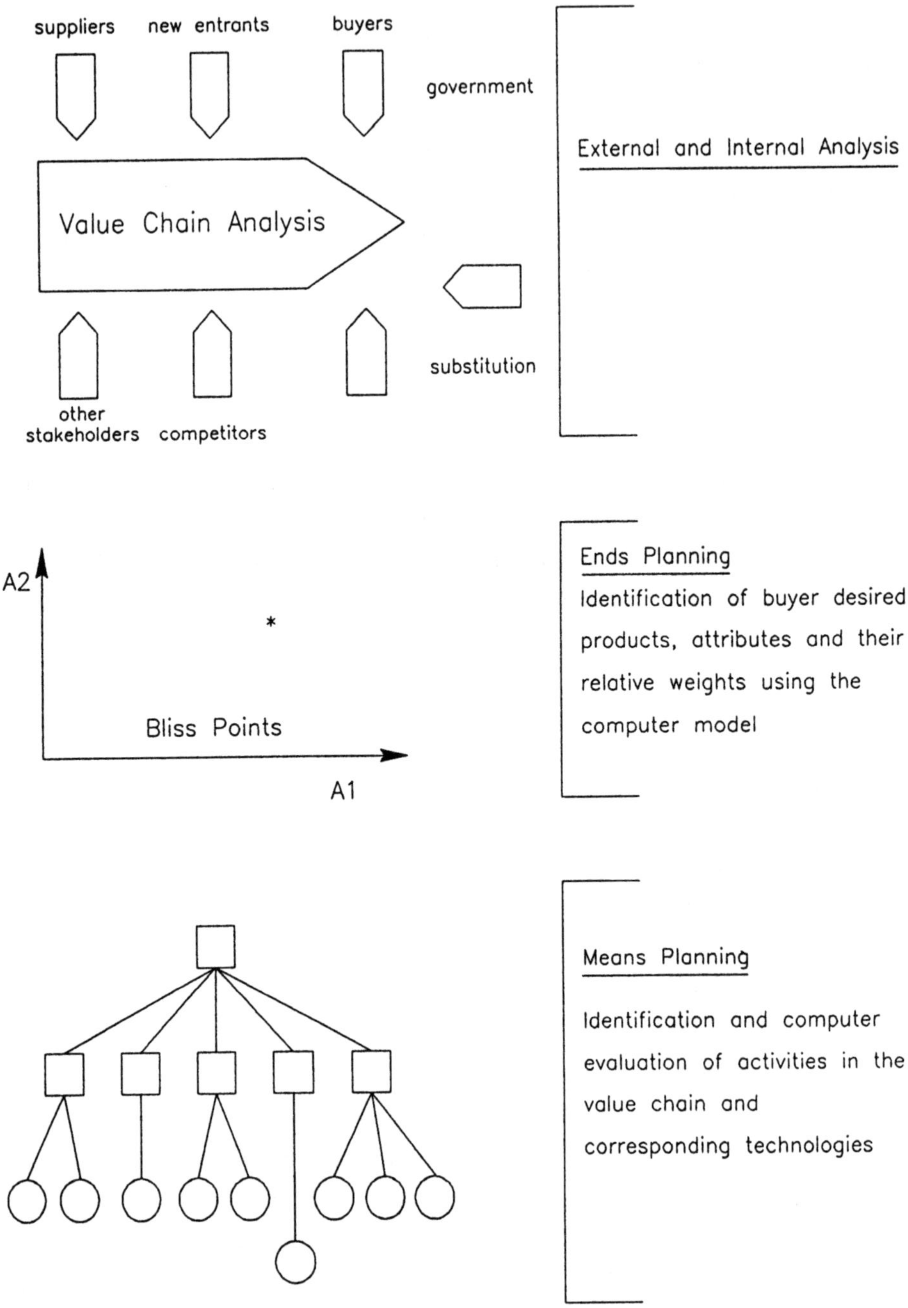

to the organization (Porter, 1985).

The second phase deals with **ends planning**. Here, the desirable characteristics of a product line are defined from a consumer or marketing perspective. Such product characteristics are then translated into operational product design and manufacturing performance objectives in terms of cost, quality, flexibility and service. The output of ends planning is a specification of desirable performance objectives to be met by available product or process technologies. The process of developing such objectives is closely related to Ackoff's idealized design process (Ackoff, 1981). In the manufacturing literature the resulting set of objectives expressed in terms of cost, quality, flexibility and dependability is often referred to as the manufacturing mission statement (Cohen and Lee, 1985). The key to effective ends planning is that it leads to understandable benchmarks for more detailed technology assessment (Kleindorfer and Partovi, 1990). Ends planning objectives might be stated in terms of improvement objectives relative to the status quo. Examples include manufacturing products which are 20 per cent cheaper or available 50 per cent faster, or products with 10 per cent less defects. It would then remain for the technology evaluation process to determine how well competing technologies might contribute to the accomplishment of these improvement objectives.

Means planning, the third phase of our TEM is directed at the detailed evaluation of specific technologies. A computer-based model to evaluate different technologies in the value added chain has been developed. This model, unlike most technological selection and justification models, considers benefits as well as cost. This model will be discussed further in the following section.

4. TECHNOLOGY EVALUATION MODEL

In this section a model for evaluating the choice of technology is proposed. This model serves as a focus for combining the results of the strategic planning and ends planning activities with technology choice as it relates to various activities along the organization's value-added chain (Porter, 1980).

The model is in the form of a hierarchy that includes the forces driving competition, their components, activities in the value chain, and corresponding technologies in those activities. It is possible to assess each level's priority in this hierarchy using the analytical hierarchy process (AHP) of Saaty (1982). Figure 2 depicts the structure of the hierarchy. The elements can be classified into three groups: strategic forces driving competitive advantage, value chain activities, and characteristics of available technologies. We first discuss these three different groups of elements in detail. Then we review the application of the AHP process based on this hierarchy.

4.1 Strategic Forces

Assuming long-run profit as the overall performance objective, strategic forces driving competition can be divided into two categories: those affecting cost and those affecting revenue. In manufacturing terms, revenue-enhancing objectives of technology are measured in terms of quality, flexibility and dependability (Wheelwright, 1984). To assess the impact of technology on cost and revenue components in detail, it is useful to operationalize these components further. Our intention here is to be illustrative. Clearly, particular firms or industries may find other performance criteria useful in operationalizing the strategic impact of technology. Indeed, developing the appropriate performance evaluation hierarchy corresponding to Figure 2 in each specific context is an important ingredient of the evaluation process envisioned here.

The operationalization of cost and revenue objectives on specific dimensions is briefly presented here. For a more detailed discussion of these objectives see Kleindorfer and Partovi (1990).

4.1.1 Cost

A firm's unit costs can be divided into two categories, direct and indirect cost.

Figure 2. Hierarchical Model Relating Manufacturing Strategy to Choice of Technology in the Value Chain

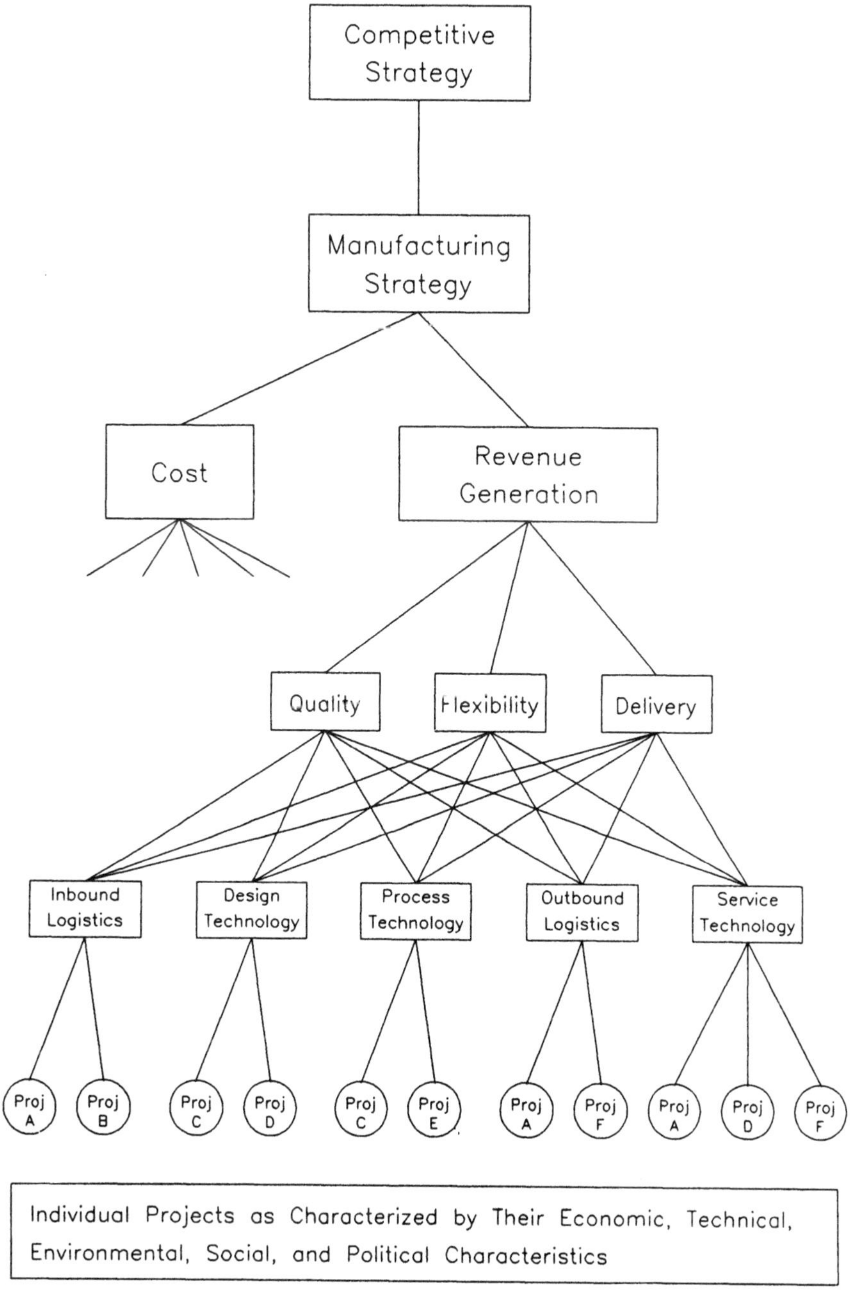

4.1.2 Quality

Product quality is emerging as perhaps the most important criterion of manufacturing success. The great success stories of Japanese manufacturing derive from the dedication of their factories to produce products according to the needs of customers and to eliminate defective products. Quality can be divided into eight sub-categories: performance, features, reliability, durability, aesthetics, conformance, serviceability, and perceived quality (Garwin, 1988).

4.1.3 Delivery

Delivery is an objective specifying constraints on time, space and volume availability of product. Dependability can be divided into the following dimensions: availability in space, availability in time, and availability in volume (Kleindorfer and Partovi, 1990).

4.1.4 Flexibility

Flexibility is the ability to respond to external and internal changes. These changes can occur with respect to consumer demand, technical innovations, the economy, or government regulations. Flexibility can be divided into the following dimensions: Product Flexibility, and Process Flexibility.

4.2 Activities of the Value Chain

The above operationalization of strategic forces should allow a specification (via ends planning) of performance objectives important for strategic advantage. In order to evaluate the contribution of specific technologies to accomplishing these objectives, it is convenient to analyze the impact of alternative technologies in terms of their consequences for specific activities in the value chain. For illustrative purposes, we consider the impact of technology on the following activities in the value chain: inbound logistics, product design, manufacturing process, outbound logistics, sales and after-sale service (Porter, 1985).

4.3 Evaluation Criteria for Specific Projects

At the lowest level in the hierarchy of Figure 2 are the specific technology projects themselves. The ultimate intent of the evaluation model is, of course, to prioritize these technologies. The question of interest is to evaluate how well a given technology contributes to a particular strategic objective (e.g., cost reduction) through specific activities in the value chain.

4.4 Application of AHP in Technology Choice

Using the proposed model, it is possible to link a particular strategy to a specific technology in the value chain. In doing so, let us reconsider Figure 2. The reader should imagine that this figure has developed through the process of formulation of the problem, ends planning and the operationalization of specific performance criteria related to technology choice (illustrated above). At this point, key decision makers will have agreed on a performance hierarchy and available technology projects. Now their task is to evaluate each available technology using this performance hierarchy. This evaluation process proceeds using established procedures (and supporting software) associated with (AHP). We only briefly describe this process here since the details are by now relatively well understood in general (Saaty, 1982).

The first level of the overall hierarchy in Figure 2 can be represented by the four forces driving competitiveness (as seen in Figure 3). Decision makers are asked to specify the relative importance of each of these forces in achieving competitive advantage. The AHP elicits the necessary information in the form of pairwise comparisons. For example, how much more important is cost than quality in achieving the manufacturing strategy? AHP uses a tested measurement scale and an analytical procedure to process these judgements (for further details see Saaty, 1982).

Figure 3. Hierarchy Relating Competitive Strategy of the Firm to Forces Driving Competition

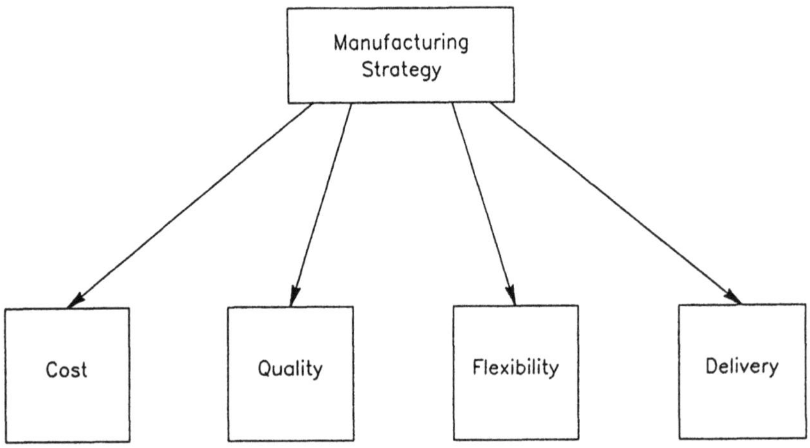

In order to obtain meaningful responses to this evaluation, the competitive forces have to be benchmarked explicitly. Such benchmarks typically follow from a strategic audit indicating desired improvement goals in cost, quality, flexibility and delivery. The results of such an audit would yield objectives such as: reduce product price/cost by 10% (cost); increase product precision by 20% (quality); decrease product lead-time by 10% (flexibility); and increase product availability by 10% (dependability). After such improvement goals have been concisely stated, the evaluation which would accompany Figure 3 would be conducted. The outcome of this evaluation is a preference matrix and corresponding set of weights for the first level of the model. These weights represent the relative importance of the forces driving competition as operationalized through the specified improvement goals or benchmark (see Kleindorfer and Partovi, 1990). Let $W = (W_1, W_2, W_3, W_4)$ be the vector of weights for the forces of competition.

Once the relative priorities of the forces driving competition have been obtained, the next step is to find the importance of particular activities in the value chain with respect to each of these forces. Referring to Figure 2, this yields four sub-hierarchies, one for each of the forces driving competition advantage. The "Flexibility Sub-Hierarchy" is illustrated in Figure 4. Here decision makers would judge, for example, the relative importance of design activities in comparison with inbound logistics in accomplishing the stated flexibility benchmark.

Figure 4. The Hierarchy of Evaluating Technologies of the Value Chain with Respect to Flexibility Strategy

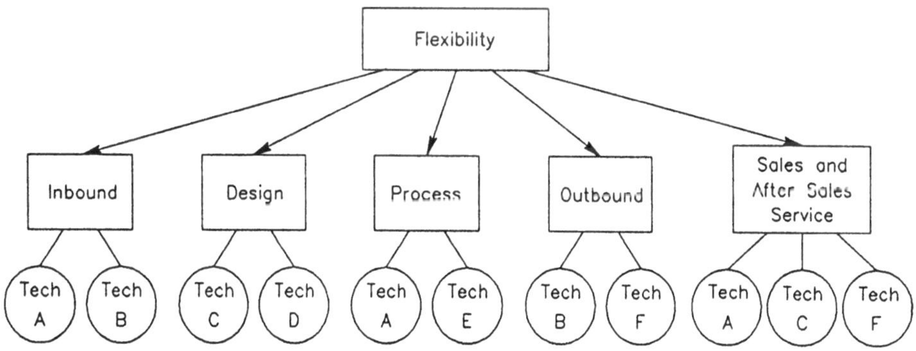

Next, the available technologies are evaluated with respect to each of the activities in the value chain. Note that these evaluations must be placed in the context of a particular strategy such as flexibility. For example, one would compare technology A with technology B in improving inbound logistics with respect to flexibility (Figure 4). The weights of alternate technologies must be aggregated across all activities to determine the overall technology weight with respect to a given competitive force (such as flexibility). This is accomplished using a weighted average approach. Consider the following example. Referring to Figure 4, assume technology A's weights with respect to Inbound, Process and

Service are .3, .5, and .4, respectively. Also assume that the weights of these three activities with respect to flexibility are .2, .1, and .3, respectively. Then technology A's weight with respect to flexibility is: (.2)(.3) + (.1)(.5) + (.3)(.4) = .23. The weights of the other technologies would be determined in a similar fashion.

Let L represent the normalized matrix of AHP priorities of available technologies in the cost, quality, flexibility and dependability sub-hierarchies (e.g., Figure 4). These priorities measure the relative contributions of specified technologies to the accomplishment of the respective cost, quality, flexibility and dependability objectives. Assuming n technology options and for competitive forces, L is the n x 4 matrix given below. Each column of L represents the weights of a given technology with respect to a specific competitive force.

$$L = \begin{matrix} L_{11} & L_{12} & L_{13} & L_{14} \\ L_{21} & L_{22} & L_{23} & L_{24} \\ \cdot & \cdot & \cdot & \cdot \\ \cdot & \cdot & \cdot & \cdot \\ L_{n1} & L_{n2} & L_{n3} & L_{n4} \end{matrix}$$

The technology weights given in L must be adjusted by the weights of the four competitive forces determined earlier (the W-vector). Define $T = (T_1, T_2, \ldots, T_n)$ to be the final technology weights or global priorities. These weights can be determined using the weighted average approach previously described. Formally, the matrix product of L and W must be computed: T = LW. This procedure accomodates the fact that different technologies may be relevant only to certain attributes on the value chain as shown in Figure 4.

In the following sections, the proposed AHP model is tested. Its adequacy in representing the relationship between the forces driving competition and choice of technology is verified, and the implementation of the methodology in a manufacturing environment is described.

5. EXPERIMENT

A laboratory experiment compared the normative model presented in the previous section with traditional and intuitive approaches to technology selection (Partovi, 1989b). The research in this experiment was directed at determining whether the use of the proposed TEM, makes any difference in the performance and behavior of the planning teams that use the model over those using traditional methods.

The experiment proceeded as follows (see Partovi, 1989a; Partovi, 1989b for details). Thirty-three teams of three subjects each solved a standard case, "The Mirrasou Vineyards" (Marshall, et. al, 1975). These subjects were part-time MBA students studying "Manufacturing and Technology". The purpose of the case analysis was to determine whether a new technology should displace the current one used by Mirrasou Vineyards for harvesting its grapes. Roughly half of the teams (16) were given no specific training on technology evaluation beyond what they had learned in their basic management and finance courses. The other half were introduced to the TEM framework described above and were encouraged to use this approach. After solving the case, each planning team evaluated its own performance. Also, a group of expert judges evaluated the case write-ups of these experimental planning teams. Basic statistical analyses were then performed to compare the performance and perceptions of those which had been exposed to the TEM framework versus those groups which had not been exposed.

The experimental results suggest that using the technology selection model: (1) may decrease communication among group members and allow one team member to control and dominate others in making group decisions; (2) requires more time than traditional justification models; (3) promotes problem segmentation; and (4) provides a more rewarding experience than intuitive procedures.

The majority of the experts indicated that the treatment group achieved higher performance levels in terms of: (1) the coherence of their reasoning; and (2) credible support for the final decision. Finally, in informal communications with the judges after the evaluations were completed, some stated that they felt that using the proposed methodology supports a more liberal attitude towards technology choice in manufacturing, an attitude lacking in more traditional technology decision-making.

6. A CASE STUDY

The author has been involved in the implementation of the proposed technological selection methodology in aiding managers to select advanced manufacturing technologies. To preserve anonymity, names are disguised, and for the sake of brevity, the particulars of the case are condensed (see Partovi, 1988, for a more detailed discussion of this case). The case provides a practical example of the TEM proposed in previous sections.

ABC Inc. is a leading manufacturer of all digital mass measurement weighing products for industrial use. It provides weight feeding equipment used in the processed food, plastic and chemical industries. ABC Inc. consists of three main divisions: marketing, engineering/quality services and manufacturing. Some background on these divisions is needed to understand how the proposed technologies will affect this company.

The marketing division consists of: sales personnel who represent ABC around the world, the test laboratory which deals with material flow characteristics required by customers, and the market research department which determines how the company should shape its marketing policy.

The engineering and quality service division consists of six different departments: product engineering, project engineering, quality assurance, services, parts and spares, and engineering management services. The product engineering department turns the latest feeding and control technology from pure research into practical hardware. Project engineering is responsible for all orders, from sizing the equipment through final testing. Quality assurance establishes quality levels for product parts and assures that those levels are achieved in production. The service department makes certain that the customer's downtime is as small as possible. In addition, service is responsible for training customers in the use of the equipment. Parts and spares operates as a small-scale sales division which handles the spare parts requirement of the customer. And finally, engineering management services coordinates engineering documents and writes product manuals.

The manufacturing function includes the operations directly concerned with the making of the product and various services to these productive operations. Manufacturing consists of two major functional departments with two additional supporting groups. The two primary functional departments are production and planning. The production department contains the sections needed to fabricate and assemble the metal parts and electronic components into finished goods. The planning department provides all materials and production schedules. The planning department is also responsible for shipping, storage, and internal transportation of the materials and products.

The first part of TEM consisted of extensive internal analysis which included a detailed understanding of value chain activities and their corresponding technologies as well as standard financial and personnel analyses. In addition, an extensive analysis of customers, competitors, suppliers, substitution, and government and other stakeholders was performed.

6.1 Idealization

The management of the ABC completed the idealization process, and reached a consensus on the following product properties which are necessary in order to sustain competitiveness.

1) To provide customers with very reliable products.
2) To provide customers with very precise and high performing machines.
3) To provide unique services that customers value and are willing to purchase.
4) To reduce the price of products to a maximum of 10% above the price of competition in order to sustain competitiveness.

Once these general competitive strategies were outlined, the management of ABC defined and benchmarked criteria which correspond to manufacturing strategy as follows.

6.1.1 Cost

This is referred to as unit cost of feeder, including labor, material, equipment, and information efficiency. Management believes that in order for ABC to restore its competitiveness, it should reduce its products' unit prices by 20%.

6.1.2 Quality

This is the most critical criterion for ABC with respect to its expected future market. Quality in this industry means very reliable and high precision equipment. The management of ABC believes that increasing the precision of feeders by 50% is realistic.

6.1.3 Delivery

This is the degree to which customers may expect immediate delivery of the equipment they have ordered. Recently, some Japanese competitors in the foreign market have obliged ABC to be more cautious about its timing of manufacturing and delivery. Ideally, ABC wants to decrease the time between getting an order and delivery of the equipment by 25%.

6.1.4 Flexibility

In this context, flexibility means the ability to meet different customers' requirements. ABC uses a job-shop manufacturing process and thus has built-in flexibility. Flexibility is thus not as critical an issue to implement as are the cost, quality and delivery criteria. Figure 5 shows the criteria and their weights are given by the management of ABC Inc.

Figure 5. Hierarchy of Forces Driving Competition and Their Corresponding Weights for ABC Inc.

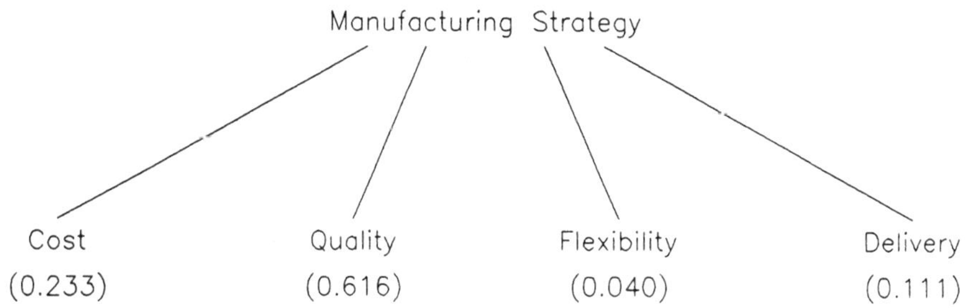

6.2 Potential Technological Alternatives in the Activity Chain

There are many technological advances that can improve the ABC's strategic performance. These technologies vary in the level of sophistication. Identification of potential technologies in the activity chain started with a brainstorming session in which the participating managers suggested new technologies for their own or other divisions. The first option suggested was a general business control system. The main function of this software, the Manufacturing, Accounting, Production, and Information Control System (MAPICS) package, is to coordinate different operations in the factory and to mesh them together. Moreover, this software can facilitate external transactions with vendors and customers. In short, MAPICS can reduce the duration of activities from start to finish. It can be useful in production planning, marketing research, and project engineering departments.

Computer Aided Design (CAD) was the second technology considered for potential investment. CAD uses geometric modeling techniques to create an unambiguous model of the parts which is used to produce the engineering drawings. Important features of CAD for ABC are the electronic design file and data base to which all departments have access. CAD also improves communications with vendors and customers. The product and project engineering departments will be the primary users of CAD. The technology suggested for the test lab department is an automated sampling system (AUSS) that measures very accurately product performance. This machine helps to optimize the design and performance of ABC products according to customers' requirements.

In the machine shop, sheet metal cutting methods and metal finishing were activities that required renovation. Three alternative technologies were considered for the cutting operations: laser cutting, machine center, and plasma cutting. Laser cutting can do a superior job in terms of quality and precision compared to any existing method. Other advantages include faster operating time and the ability to machine areas not readily accessible by other instruments. Its drawback is cost. The major advantage of a machine center is the ability to perform a multiplicity of operations in only one setup. This eliminates the need for a number of individual machine tools, thus reducing capital equipment and labor requirements. Additional savings also result from reduced material handling, work-in-process inventory and fixture costs. A machine center also maintains consistency, resulting in higher quality parts as well as reduced scrap and inspection costs. Estimates of increased productivity with this machine ranges from 300% to 500% (Dallas, 1984). Finally, the principal advantage of the plasma arc cutting process is that it is almost equally effective on any metal regardless of hardness or refractory nature. Other advantages include the quality and speed of the cuts. The principal disadvantage is the metallurgical alteration of the surface.

Final inspection of parts and products is an important activity for improving the precision of ABC products. Using a coordinate measuring machine (CMM) is an option for technological improvement. This machine has a variety of measuring capabilities along the X, Y, and Z axes that can be used for checking location after machining or similar measurements. It can also be used as a layout machine before machining. The most important assets of this machine are its accuracy and fast response for doing 100% inspection, which complements other advanced manufacturing technologies.

There were other technology options under consideration during the managers' brainstorming session. These included turret punch press, computer controlled lathes, electro-polishing, printed circuit part locator, and automatic test equipment. For brevity, we will not consider these technologies here (see Partovi, 1988 for more details).

The following hierarchies (Figures 6-9) compare the importance of the activities and technologies in the value chain activities for each dimension of the manufacturing strategy. It should be noted that activities which are not important for a particular strategy have been omitted. Furthermore, the hierarchies are "incomplete" (Harker and Vargas, 1987; Saaty, 1986); that is, different technologies are not relevant to all activities in the value chain. The prioritization of technologies for each of the cost, quality, flexibility and delivery hierarchies is also shown in each figure.

Once the technologies were prioritized for each manufacturing strategy, the results were combined and integrated using the strategies' priorities. The column data are obtained from Figures 6-9, while the strategy weights, shown as column multipliers, were obtained from Figure 5.

$$\begin{matrix} W_{CAD} \\ W_{AUSS} \\ W_{Laser} \\ W_{MA.CTR} \\ W_{Plasma} \\ W_{CMM} \\ W_{MAPICS} \end{matrix} = (0.233) \begin{matrix} \underline{\text{COST}} \\ 0.000 \\ 0.000 \\ 0.194 \\ 0.103 \\ 0.036 \\ 0.000 \\ 0.667 \end{matrix} + (0.616) \begin{matrix} \underline{\text{QUALITY}} \\ 0.400 \\ 0.200 \\ 0.114 \\ 0.057 \\ 0.029 \\ 0.200 \\ 0.000 \end{matrix} + (0.040) \begin{matrix} \underline{\text{FLEXIBILITY}} \\ 0.667 \\ 0.333 \\ 0.000 \\ 0.000 \\ 0.000 \\ 0.000 \\ 0.000 \end{matrix}$$

$$+ (0.111) \begin{matrix} \underline{\text{DELIVERY}} \\ 0.091 \\ 0.000 \\ 0.064 \\ 0.042 \\ 0.016 \\ 0.000 \\ 0.787 \end{matrix} = \begin{matrix} 0.284 \\ 0.138 \\ 0.124 \\ 0.059 \\ 0.027 \\ 0.124 \\ 0.244 \end{matrix}$$

Figure 6. The Quality Hierarchy with Priorities for Activities and Technologies

Technology	Priorities
CAD	0.400
AUSS	0.200
LASER	0.114
MACH CTR	0.057
PLASMA	0.029
CMM	0.200

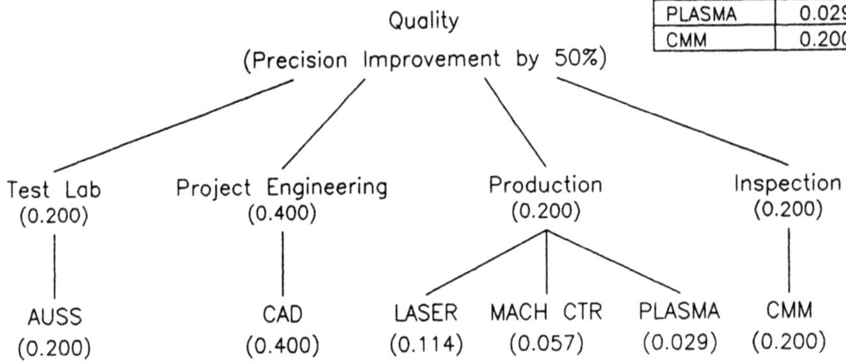

Figure 7. Reducing Cost of the Product by 20% with Priorities for Activities and Technologies

Technology	Priorities
MAPICS	0.667
LASER	0.194
MACH CTR	0.103
PLASMA	0.036

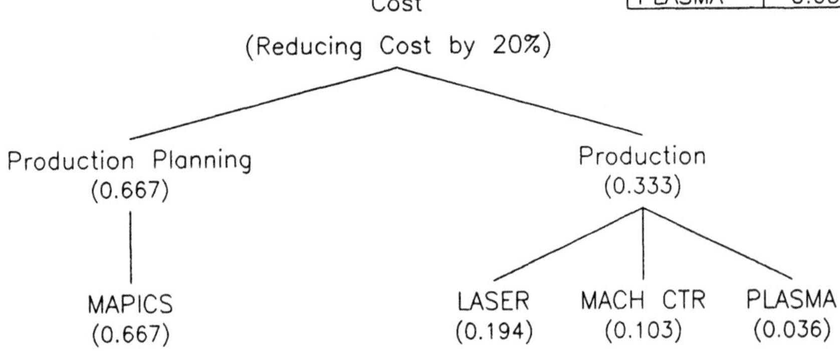

Figure 8. The Flexibility Hierarchy with Priorities for Activities and Technologies

Technology	Priorities
CAD	0.667
AUSS	0.333

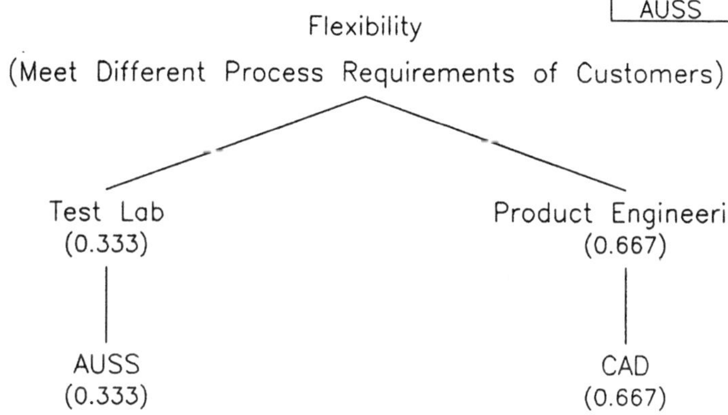

Figure 9. The Delivery Hierarchy with Priorities for Activities and Technologies

Technology	Priorities
MAPICS	0.787
CAD	0.091
LASER	0.064
MACH CTR	0.042
PLASMA	0.016

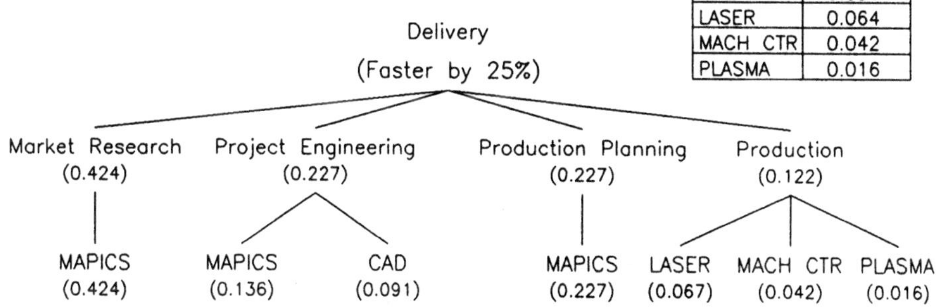

The results show the strategic assessment of competing technologies, with CAD and then MAPICS having the highest weights. It should be noted that the priorities derived by TEM are strategic in nature and are to be used jointly with standard financial measures (such as NPV) in making overall priority choices.

7. SUMMARY AND CONCLUSION

The purpose of this paper has been threefold: first, to present a methodology and corresponding model relating operations strategy to choice of technology; second, to test the proposed model for adequacy and implementability; and third, to implement the proposed methodology in a manufacturing environment.

The technology evaluation methodology presented is based loosely on Ackoff's interactive planning, and incorporates Saaty's AHP, which converts manufacturing strategies to process-distinctive features. The evaluation of the proposed methodology using laboratory experiments suggests that the use of this methodology improves the quality of decisions in terms of coherent reasoning, increased support for the final decision, and a more liberal attitude towards adoption of advanced manufacturing technologies. However, the model may decrease communication within the technology planning group. Finally, the proposed methodology was implemented in a manufacturing environment. As part of the implementation, the forces that shape the firm's competitiveness, operationalization of variables used in the model, and the characteristics of technologies were studied.

Several interesting questions are raised by this methodology and warrant further research. First, while financial criteria are typically used as screens for feasible technologies, in actuality trade-offs exist among the strategic dimensions and financial dimensions. Therefore, an explicit and common treatment of financial and strategic dimensions within a unified framework would be highly desirable. Second, the technological evaluation suggested in this research can take an opposing view. That is, instead of comparing given technologies with respect to a particular strategy, one can pose the question of where do we <u>need</u>

advanced technology in the value chain, and what characteristics are required in that technology.

ACKNOWLEDGEMENT

Appreciation is extended to Professor Kleindorfer for his comments on this article.

8. REFERENCES

Ackoff, Russell L. (1981). **Creating the Corporate Future.** John Wiley & Sons.

Blois, K. J. (1986). "Manufacturing Technology as a Competitive Weapon," **Long Range Planning, 19** (4), 63-70.

Cohen, Morris A. and Hau L. Lee. (1985). "Manufacturing Strategy Concepts and Methods," **Management of Productivity and Technology in Manufacturing**. Paul R. Kleindorfer (ed.). New York: Plenum Press.

Dallas, Daniel B. (1984). **Tool and Manufacturing Engineers Handbook.** Society of Manufacturing Engineers.

Diaz, Andres E. (1986). "The Software Portfolio: Priority Assignment Tool Provide Basic for Resource Allocation," **Industrial Engineering, 18** (3), 58-65.

Freeman, R. Edward (1983). "Managing Strategic Challenge in Telecommunications," **Columbia Journal of World Business, 18** (1), 8-18

Garwin, D.A. (1988). **Managing Quality.** New York: Free Press.

Harker, P. T. and L. G. Vargas. (1987). "The Theory of Ratio Scale Estimation: Saaty's Analytic Hierarchy Process," **Management Science, 33** (11), 1383-1403.

Jelinek, Mariam and Joel D. Golhar. (1983). "The Interface between Strategy and Manufacturing Technology," **Columbia Journal of World Business, 18** (1), 26-36.

Kaplan, Robert S. (1986). "Accounting Lag: The Obsolescence of Cost Accounting Systems," **California Management Review, 28** (2), 174-199.

Kleindorfer, Paul R. and Fariborz Y. Partovi. (1990). "Integrating Manufacturing Strategy and Technology Choice," **European Journal of Operations Research**, (in press).

Marshall, Paul W., William J. Abernathy, Jeffrey G. Miller, Richard P. Olsen, and Richard S. Rosenbloom. (1975). **Operations Management: Text and Cases.** Homewood, Illinois: R. D Irwin.

Nobel, Jean L. (1989). "Techniques for Cost Justifying CIM," **The Journal of Business Strategy, 10** (1), 44-49.

Park, Chan S. and Young K. Son (1989). "An Economic Evaluation Model for Advanced Manufacturing Systems," **The Engineering Economist, 34** (1), 1-26.

Partovi, Fariborz Y. (1988). "Integrating Manufacturing Strategy and Technology Choice," Ph.D. Dissertation, University of Pennsylvania.

Partovi, Fariborz Y. (1989a). "An Experimental Study of Computer Support Systems for Strategic Choice of Technology," **Northeast DSI Proceedings.** Baltimore, MD.

Partovi, Fariborz Y. (1989b). "A Comparative Study of a Computer Aided Evaluation Method and Ad Hoc Method of Strategic Choice of Technology," **Working Paper Series, Department of Management and Organizational Sciences**, Drexel University.

Porter, Michael E. (1980). **Competitive Strategy.** New York: Free Press.

Porter, Michael E. (1985). **Competitive Advantage.** New York: Free Press.

Saaty, Thomas L. (1982). **Decision Making for Leaders.** Belmont, CA: Lifetime Publications.

Saaty, Thomas L. (1986). "Axiomatic Foundation of the Analytic Hierarchy Process," **Management Science, 32** (7), 841-855.

Skinner, Wickham (1985). **Manufacturing: The Formidable Competitive Weapon.** New York: John Wiley & Sons.

Stevens, Kathy C. and L. R. Martin (1989). "Could Internal Decision Models Be to Blame," **Advanced Management Journal, 45** (1), 32-36.

Swamidass, Paul M. (1987). "Planning for Manufacturing Technology," **Long Range Planning, 20** (5), 125-133.

Swindle, R. (1985). "Financial Justification of Capital Projects," **Proceedings of the Autofact '85 Conference**, Detroit, MI, 5.1 - 5.17.

Wheelwright, Steven C. (1984). "Manufacturing Strategy: Defining the Missing Link," **Strategic Management Journal, 5** (1), 77-91.

THE DANTE MODEL: DYNAMIC APPRAISAL OF NETWORK TECHNOLOGIES AND EQUIPMENT

Nicolas V. Danila
Professor at Groupe Centre de Perfectionnement aux Affairs
Managing Director, N. V. D. Consultants
108, bd Malesherbes 75017 Paris

ABSTRACT

The formulation and evaluation of advanced manufacturing technology programs are important activities within the strategic planning process. This paper describes the seven steps of the DANTE (Dynamic Appraisal of Network Technologies and Equipment) Model which has been developed to support these activities. The DANTE Model utilizes analytical and intuitive modelling approaches, and incorporates quantitative as well as qualitative data. This approach allows dependency relationships between individual projects, and defines a network of technologies and equipment which constitutes a complete investment program. A case study describing the application of the DANTE Model to the formulation and evaluation of investment options for a large-scale automation program is presented. A discussion of the strengths and limitations of the model are also given.

1. INTRODUCTION

The intensified level of global competition must be reflected in the development of technological strategies which support the corporate strategy. The rapid rate of technological change itself requires managers to initiate new technology investment programs in order to keep pace with the competition. Unfortunately, the costs associated with these programs are escalating rapidly. The programs themselves are increasing in complexity and must often be represented as a network of technology-related projects with certain dependency relationships. Clearly, evaluation of such technology networks and appraisal of their risks is crucial to the firm's survival and must be linked to corporate strategy (Andrews 1980).

An interdisciplinary technique, such as the support graph, seems to be particularly effective and efficient for the formulation stage (Danila 1979). The support graph links objectives to activities, and, in turn, links the technological, human and financial resources to these activities. There is a large literature concerning program management and evaluation (e.g., Kerzner 1984, Archibald 1976). However, few authors propose an evaluation method for dependent projects and a global evaluation of technology networks, whose configurations are often changing over time. The evolutionary nature of such programs is quite common and contribute to difficulties in evaluation. Following the evaluation phase, these technology programs require organization, planning, control and finally implementation. However, success in these subsequent activities depends on the quality of the formulation and evaluation of the technology network.

About ten years ago a new approach called the DANTE (Dynamic Appraisal of Network Technologies and Equipment) Model was developed to address the formulation and evaluation of dependent technology networks. This method first defines the frontier of a network of technologies and equipment, and then aggregates the quantitative and qualitative criteria used during the evaluation process.

The next section of the paper presents the objectives of the model and potential areas of application. This is followed by a description of the seven steps of the model. Next, a case study is presented, and the application of the seven steps of the model are described. The following section discusses model application issues including its linkage with strategic planning, institutional learning, creativity issues and limits of the model. The final section offers some conclusions.

2. MODEL DESCRIPTION

The DANTE Model utilizes quantitative financial evaluation criteria such as Net Present Value (NPV), Internal Rate of Return

(IRR) and Differential Appreciation Model (DAM), which will be described later. It also draws upon qualitative approaches such as the support graph technique (Danila 1989), brainstorming or Delphi methods (Helmer 1966), and an integrative methodology such as the multicriteria approach called Electre-Oreste (Danila 1979). The three principal objectives of the DANTE Model are:

1. Define each technology investment option as a network of technologies and equipment. This approach is particularly useful when different projects and/or types of equipment constitute the technology investment program. The structuring of all possible configurations and the collection of data should be accomplished within clearly defined boundaries.

2. Assist the decision-making process by including all the important criteria, whether quantitative or qualitative in nature. The aggregation of these criteria should closely represent the preference function of the decision-makers.

3. Propose and help implement the conditions required for the success of the best technology network configuration after checking the feasibility of different solutions.

The areas of potential application of the DANTE Model for the evaluation of Advanced Manufacturing Technology Networks are:

1. Automation process programs such as motor car production and telecommunication stations;

2. Complex pilot plants in fields such as biotechnology, aerospace, and chemicals; and

3. Development and engineering technology programs in chemistry, textiles, computers and any other sectors where links between projects which are part of the same program must be considered within the evaluation process.

2.1 The seven Steps of the DANTE Model

The seven steps of the DANTE Model are as follows:

STEP 1: IDENTIFICATION

Identify the problem context, objectives, technologies and equipment, and resources needed. Use creativity, imagination and intuition to help capture all the necessary information.

STEP 2: CONFIGURATIONS

Construct all possible configurations of technology investment programs using creative techniques such as the support graph. Structuring the program information during this step must be accomplished without losing sight of the supporting quantitative and qualitative data.

STEP 3: CRITERIA

List all of the evaluation criteria, and categorize them as quantitative and qualitative. A good balance between both types should exist (Danila 1985).

STEP 4: EVALUATION

Calculate the values of the quantitative criteria selected and describe the dimensions of the qualitative criteria for all of the configurations requiring evaluation (Danila 1983).

STEP 5: AGGREGATION

Aggregate the various criteria using different weights for each, and determine the overall weights for the various technology programs. Rank the various technology programs considered (Danila 1988a).

STEP 6: NEGOTIATION

Analyze the rankings from STEP 5 above. Examine the feasibility of the technology programs with respect to the objectives to be achieved and the resources available. The rankings found during STEP 5 can be modified by this analysis.

STEP 7: ACTION

Select a solution to be implemented and an action plan which will ensure the realization of all benefits from the selected configuration (Danila 1988b).

Figure 1 presents the seven steps of the DANTE Model.

Figure 1. The Seven Steps of the DANTE Model

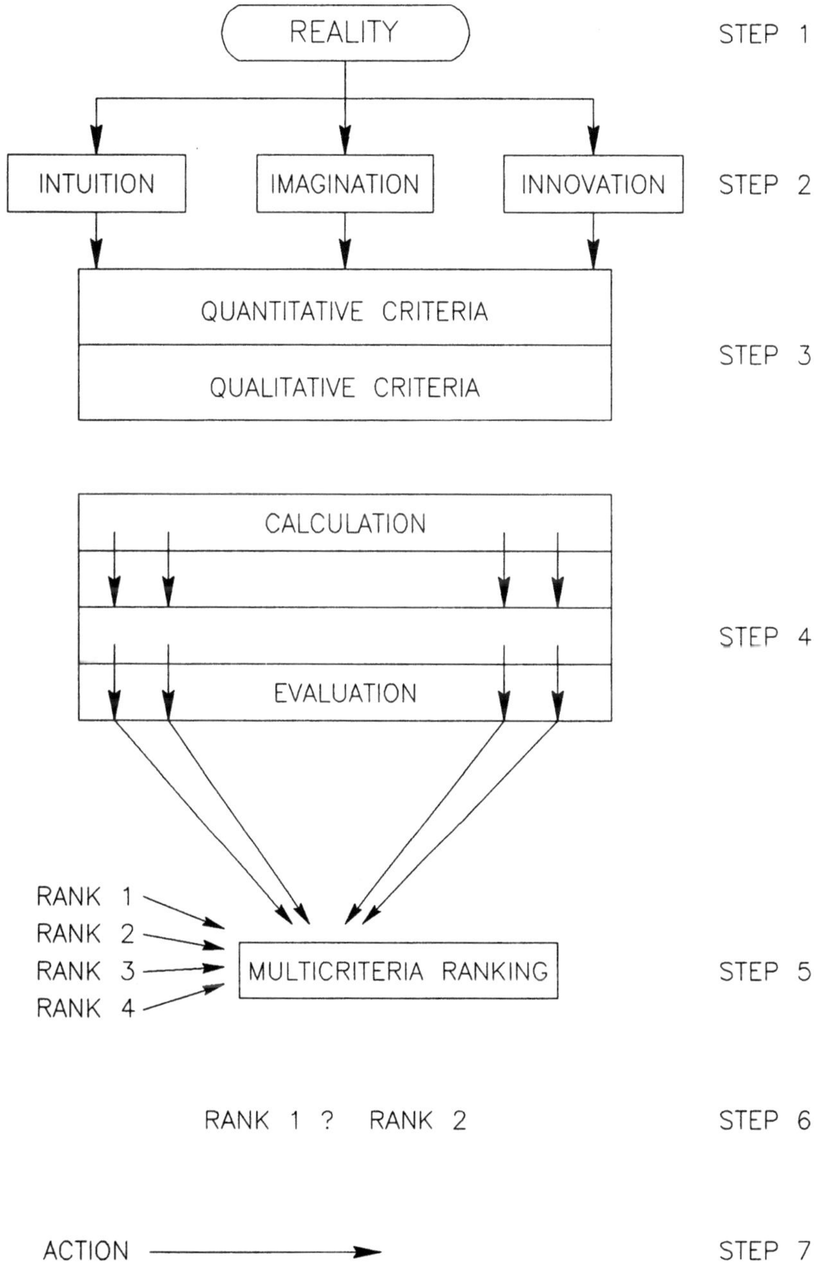

3. CASE STUDY

3.1 Background

During the 1970's a major technology program was proposed which would replace a manual processing system with an automatic one. This National Technology Program would affect all French Government Departments and several thousand employees. The estimated total cost of this program was 2 billion U. S. dollars, and the project lifetime was estimated as 26 years. The program contains a set of smaller projects, all of which are distinct but interdependent. To avoid the "π" rule (the final cost of a technology program is the estimated cost multiplied by 3.14), the organization conducted a detailed evaluation and attempted to exercise management and control over the program.

It is important to note that the proposed automation program did not address all of the operational processes, but only two of them. Figure 2 shows where the technology investments will be incorporated into the manual system. Specifically, the automatic system will address the links B-C and D-E. Prior to evaluation, it was necessary to define the technology network, the type of equipment, and the resulting cash payments and personal expenses.

Figure 2. Total Process System Configuration

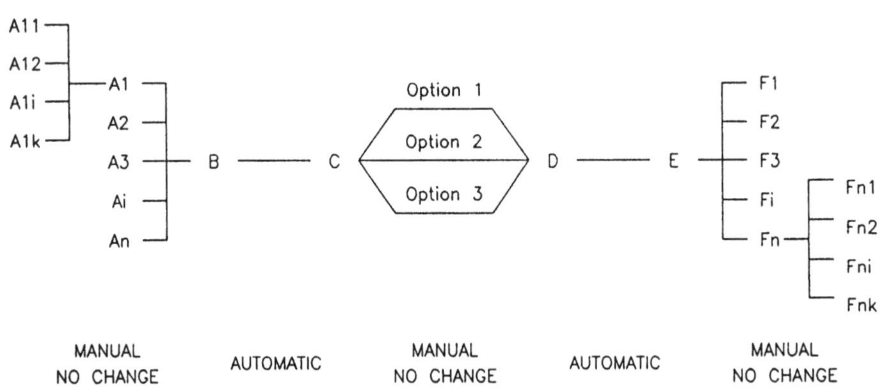

In what follows we present the application of the DANTE Model, step by step. After identifying the possible technology options, it was necessary to identify and aggregate the various criteria, and check the feasibility of the complex technological networks. The DANTE Model was used to address these difficulties by helping to select the best configuration and by assisting in its implementation.

3.2 Application

3.2.1 Step 1: Identification

During this step, the objectives of the proposed automation systems were described. The list of potential technologies and types of equipment were developed, and estimates of the total amount of resources required were made. The lifetime of the overall program was checked to determine if it matched with the lifetime of the equipment.

Three classes of equipment were identified, and are defined as follows:

CLASS A: A1, A2, A3;
CLASS B: B1, B2; and
CLASS C: C1, C2, C3, C4, C5, C6.

Each configuration requires at least one type of equipment from each class.

For all classes of equipment, expenses can be grouped into five categories. The distribution of expenses by category is as follows:

Purchase price...........68%
setup....................10%
adjustments..............11%
development...............9%
training courses..........2%

3.2.2 Configurations

There were a variety of potential combinations of options and for that reason each of the 53 different subprojects needed its own configuration. A subproject contains a building, a technological configuration and human resources, designed to meet the demand. It should be noted that it was impossible to build two similar configurations for a given subproject. Figure 3 depicts three typical configurations.

Figure 3. Different Configurations of Technology Equipment

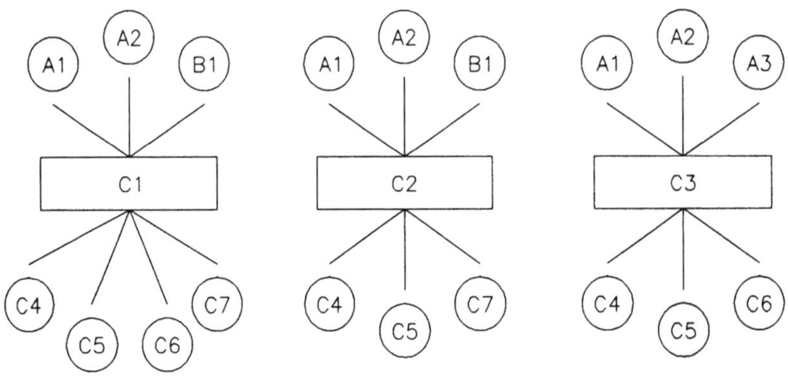

For each configuration, construction investment also differs to some extent. However, it was found that there were in fact only a dozen classes of different configurations, depending upon the level of manufacturing activities. Also, it was necessary to have a minimum capacity for each of the three types of equipment (A, B and C). All of the configurations were evaluated through the use of scenarios.

3.2.3 Criteria

Quantitative

Three financial criteria were initially considered: Net Present Value (NPV), Internal Rate of Return (IRR) and Differential Appreciation Model (DAM). NPV requires the selection of a discount rate and needs the estimation of cash inflows. Although IRR does not require the selection of a discount rate, it too needs an estimate of cash inflows. Since this project was performed for the French government, the measurement of cash inflows was very difficult to evaluate and almost meaningless. As a result, NPV and IRR were not used. The DAM seemed more adaptable to the information available and was used to calculate the discounted expenses for the program.

Qualitative

The selection of qualitative criteria was linked to a wide range of managerial activities, and considered strategic, commercial, human, environmental, technological and social factors. Formally, six criteria were proposed:

1. The necessity of meeting demand (market pull);
2. the regularity and quality of the service provided by the proposed system (environment);
3. the possibility of exporting this technology (strategy and industrial policy);
4. the vulnerability of the system and the resulting consequences of failure (strategy);
5. the industrial development of some specific industrial sectors affected by the proposed system, including electronics, robotics and computers (technology); and
6. the working conditions of the people associated with the proposed system (human resources).

Some of the above criteria had specific sub-criteria. For example, the working conditions subcriteria were:

1. the importance of scheduling evening work;

2. the replacement of old and obsolete buildings and equipment;

3. the number of routine, elementary operations eliminated;

4. the required change in the job characteristics to allow job enrichment;

5. the flexibility of working hours; and

6. the extent to which power is delegated.

The criteria used were heterogenous and their influence on decision-making is very specific (Danila 1985).

3.2.4 Calculation

For the qualitative criteria, a four-point Likert scale was used: very good, good, poor, and very poor. A five-point scale was avoided because very often a medium evaluation is given.

The financial criteria was computed using the following formulas:

$$D = \sum_{i=1}^{5} D_{it} \tag{1}$$

$$D_{it} = \sum_{t=t_0}^{T} [A_{it}\prod_{j=1}^{t}(1+x_{ij}) - B_{it}\prod_{j=1}^{t}(1+Y_{ij})]/(1+a)^t \tag{2}$$

where:

A_{it} = expenditure of type i in constant French francs for the automatic program in year t;

B_{it} = expenditure of type i in constant French francs for the manual program in year t;

X_{it} = the differential inflation rate for expenditures of type i for the automatic program in year t, as compared to the average inflation rate;

Y_{it} = the differential inflation rate for expenditures of type i for the manual program in year t, as compared to the average inflation rate;

t_0 = the initial year of the project for discounting purposes;

T = the lifetime of the technology program;

a = the discount rate; and

the index i takes on the values of:

1 for equipment investments,

2 for buildings investments,

3 for operating expenses associated with the equipment investments,

4 for operating expenses associated with the buildings investments, and

5 for human resource expenses.

Note that in (2) above,

$$\prod_{j=1}^{t}(1+X_{ij}) = (1+X_{i1})(1+X_{i2})\ldots(1+X_{it})$$

For the equipment investment computation (equation (2) above with i = 1), the index for t ranges from -5 to 20, since year 6 is the reference year for discounting purposes, and the lifetime of the program is 26 years. It would be possible to develop different values of X_{1t} for each specific piece of equipment. In our case, we have assumed that all the equipment which is part of the automatic version has the same value of X_{1t}.

For buildings investments (equation (2) above with i = 2), we can replace A_{2t} and B_{2t} with terms which are based on standard space cost rates per square meter (m^2). Specifically, let:

$A_{2t} = S_{at} \times C_{at}$, and

$B_{2t} = S_{bt} \times C_{bt}$, where

S_{at} and S_{bt} are space requirements in m^2 for the automatic and manual versions, respectively, and C_{at} and C_{bt} are the corresponding cost rates, respectively. If in fact the inflation rates X_{2t} and Y_{2t} are the same, the formula for buildings investment can be simplified to:

$$D_{2t} = \sum_{t=-5}^{7}[(S_{at}C_{at}-S_{bt}C_{bt})\prod_{j=-5}^{t}(1+X_{2j})]/(1+a)^{t} \qquad (3)$$

where t ranges from -5 to 7 since the building program lasted 13 years.

Equation (3) can be simplified further as follows. Let:

$$S_{at} = S_{bt} + \Delta S_{t} \qquad (4)$$

and

$$C_{at} = C_{bt} + \Delta P_{t} \qquad (5)$$

where ΔS_t and ΔP_t are the space and price differentials, respectively, of the automatic version as compared to the manual version. If we replace S_{at} in (3) by the expression in (4) above, and then apply (5) we obtain:

$$D_{2t} = \sum_{t=-5}^{7}[(S_{bt}\Delta C_{t}+C_{at}\Delta S_{t})\prod_{j=-5}^{t}(1+X_{2j})]/(1+a)^{t} \qquad (6)$$

Consider the following example which illustrates the calculation of the differential flow for building investment. Assume:

$S_{bt} = 0.83\ S_{at}$ $\qquad \Delta C_{t} = 2260 - 2010 = 250$

$\Delta S_{t} = 0.17\ S_{at}$ $\qquad D_{2o} = (.83S_{ao})(250)+(2260)(.17S_{ao}) = 591.7\ S_{ao}$

Table 1 presents the differential flow for the building investment which must then be discounted. It is based on the data from all the subprojects.

The differential flow of operating expenditures $D_{3,t}$ for equipment includes maintenance, spare parts, small tools, and consumable goods. The formula is the same as for equipment investments as described above.

Table 1. Differential Flow for the Building Investments

Years	Differential Flow (Thousands of F)
1	22,653
2	12,804
3	22,756
4	71,909
5	41,842
6	26,712
7	43,187
8	49,685
9	47,094
10	15,600
11	5,913
12	11,713
13	5,271

The differential flow of operating expenditures for the buildings (i = 4) is calculated as follows. Let O_{at} and O_{bt} are the operating expenditures for the automatic and manual versions, and S_{at} and S_{bt} are defined as before. As in the building investment computations, define $O_{at} = O_{bt} + \Delta O_t$ and $S_{at} = S_{bt}+\Delta S_t$. We obtain:

$$D_{4t} = \sum_{t=-5}^{20} [(S_{bt}\Delta O_t + O_{at}\Delta S_t)\prod_{j=-5}^{7}(1+X_{4j})]/(1+a)^t \qquad (7)$$

For the reference year, the data are: $O_a=213$, $O_b=85$, $S_{bt}=0.83S_{at}$. The equation above becomes:

$$D_{4,o} = (.83)(S_{ao})(128) + (213)(.17)(S_{ao}) = 142\ S_{ao}.$$

Table 2 presents the differential flows for the building operating expenditures, computed in Table 1. Note that from 1985 onward, the space requirements were unchanged and so were the expenditures.

The differential flow in human expenditures was the most important financial factor because the number of jobs associated with the automatic program changed considerably. It concerns the global cost of all human resources.

Table 2. Differential Flow for the Operating Expenditures in Building

Years	Total Utility Surface for the Automatic Option (M2)	Differential Flows
1	3745	625
2	27955	4668
3	27955	4668
4	49219	8220
5	120913	20192
6	215877	36051
7	284262	47472
8	327617	54712
9	404782	67599
10	497937	83155
11	507737	84458
12	530937	88666
13	548293	91565
14	548293	91565

3.2.5 Step 5: Partial Aggregation

During this step a partial aggregation was used to compare the manual and automatic options. Several configurations of the automatic version were adopted and different rankings were obtained using the Electre-Oreste Model.

Electre-Oreste Model

The Electre-Oreste Model enables the aggregation of multiple criteria both quantitative and qualitative. This model is based upon the concept of an <u>outranking relation</u> defined by Roy (1985). Some definitions are needed before an outranking relation can be defined.

Let A be a finite set of feasible technology projects and assume a_j and a_k are elements of A. Let I be the set of criteria used to evaluate elements of A and let i be an element of I. Define a value function $V_i(a_j)$ which represents the value of program a_j on the ith criterion. A binary relation S_a defined on A is called an outranking relation if for a_j and $a_k \epsilon A$:

1. $a_j \; S_a \; a_k$ (a_j outranks a_k) means that, when the decision-maker's known preferences, the values of the projects with respect to the criteria, and the nature of the problem are taken into account, we have reasons to support the hypothesis that a_j is preferred to a_k (and no good reasons to reject it);
2. not $a_j \; S_a \; a_k$ means that, when the decision-maker's known preferences, the value of the projects with respect to the criteria, and the nature of the problem are taken into account, we do not have enough good reasons to support the hypothesis that a_j is preferred to a_k (but perhaps have no good reasons to reject it).

According to this definition, an outranking relation S_a must not be considered as an explanation of all preferences but only those which it is possible to establish without forcing the evaluation. This is an important advantage of this method. The outranking relation itself is based on satisfying both a concord and a non-discord test.

Concord Test

Let:

$$
\begin{aligned}
I^{+}(a_j,a_k) &= \{i \epsilon I \,|\, V_i(a_j) > V_i(a_k)\} \\
I^{=}(a_j,a_k) &= \{i \epsilon I \,|\, V_i(a_j) = V_i(a_k)\} \qquad (8) \\
I^{-}(a_j,a_k) &= \{i \epsilon I \,|\, V_i(a_j) < V_i(a_k)\}
\end{aligned}
$$

For example, $I^{+}(a_j,a_k)$ is the set consisting of all criteria for which a_j is strictly preferred to a_k, and the other sets can be described in an analogous fashion. The concord test checks the relative importance of these three sets with the hypothesis that a_j outranks a_k.

When the various criteria evaluated address a sufficiently large subset of decision problem's characteristics, the following weights can be defined:

$$P^{+}(a_j,a_k) = \sum_{i \in I^{+}(a_k,a_j)} P_i$$

$$P^{=}(a_j,a_k) = \sum_{i \in I^{=}(a_k,a_j)} P_i \qquad (9)$$

$$P^{-}(a_j,a_k) = \sum_{i \in I^{-}(a_k,a_j)} P_i$$

where P_i is weight corresponding to the importance of the ith criterion. Then the concord test may be rewritten as:

$$[P^{+}(a_j,a_k)+P^{=}(a_j,a_k)] \;/ \sum_{i \in I} P_i \;\geq\; c \qquad (10)$$

and

$$P^{+}(a_j,a_k) \;/\; P^{-}(a_j,a_k) \;\geq\; 1 \qquad (11)$$

where c is a parameter representing the minimum level of acceptable concordance, and the denominator in (10) is the sum of the weights for the P^{+}, P^{-}, and $P^{=}$ sets given in (9) above.

Non-Discord Test

The concord test described above evaluates the weights of the criteria in $I^{+}(a_j,a_k)$. The non-discord test evaluates the distance between the value function for the criteria in $I^{-}(a_j,a_k)$. Formally, for $i \epsilon I$, let D_i be called the discordance set. Let $e,f \epsilon E_i$ where $V_i(a_j) = e < V_i(a_k) = f$. A couple (e,f) is an element of D_i contained in $E_i \times E_i$ when the hypothesis $a_j \; S_a \; a_j$ is not admissable. For a_j and a_k, the non-discord test implies:

$$(V_i(a_j),\; V_i(a_k)) \not\in D_i \qquad \text{for all } i \epsilon I^{-}(a_j,a_k).$$

Of course, we one can more or less strictly define the discord couples, and instead of only one set D_i, several imbricated (i.e., overlapping) sets can be considered. We say that a_j outranks a_k if equations (10) and (11) and

$$V_i(a_k)-V_i(a_j) \leq h, \text{ for all } i \in I^-(a_j,a_k) \qquad (12)$$

for some acceptable value of h.

The Electre Model as defined above works well for independent technology projects. For dependent projects a modification of the approach is required. Networks of dependent projects are defined so that the networks themselves can be evaluated independently. The revised method, called Electre-Oreste can build new outranking relations with the same formula (Danila 1979).

For each scenario investigated (described below) the DAM was positive, indicating that financial profitability had not been obtained. The multicriteria evaluation showed that for almost all of the qualitative criteria the automatic option was preferred to the manual option. Several questions needed to be addressed at this point. Was it possible to improve the financial profitability? Was it possible to adjust the ranking and propose some different configurations?

3.2.6 Step 6: Negotiation

It was felt that multiple scenarios should be constructed and the calculations performed for each. As a result, four scenarios were evaluated: optimistic, reasonable, neutral, and pessimistic. The round square technique, a tool built to help the participants during the negotiation process, was used to discuss the four scenarios. Given some hypotheses concerning the network technologies and equipment, the scenarios proposed different configurations for the evaluation.

3.2.7 Step 7: Action

During this stage, all of the configurations were reexamined and only 45 of the original 53 were kept. The revised calculations showed that a financial equilibrium between the expenditures for the automatic and manual programs was obtained for some of the scenarios. The total savings estimated by the DANTE model was more than 13% of the total cost of the automatic program (in excess of 300 million U.S. dollars). The activities which had to be completed were enumerated and some indicators were proposed to help track the implementation. Table 3 presents all the analytical and intuitive tools used during the seven steps of the DANTE Model.

Table 3.

DANTE	ANALYTICAL TOOLS	INTUITIVE TOOLS
STEP 1: IDENTIFICATION	CHECKLIST of: - Objectives - Technologies - Equipments - Resources	BRAIN STORMING DELPHI
STEP 2: CONFIGURATION	SUPPORT GRAPH	DELPHI SUPPORT GRAPH
STEP 3: DICHOTOMY OF CRITERIA	QUANTITATIVE CRITERIA	QUALITATIVE CRITERIA
STEP 4: EVALUATION	CALCULATION	APPRAISAL
STEP 5: PARTIAL AGGREGATION	ELECTRE	ORESTE
STEP 6: NEGOTIATION	RATIONAL ANALYSIS	IMAGINATION OF NEW SOLUTIONS (ROUND SQUARE TECHNIQUE)
STEP 7: ACTION	IMPLEMENTATION	ALADIN APPROACH (Second Part)

4. DISCUSSION

4.1 Links with Strategy

Often technology projects and programs are thought of as independent entities and are managed accordingly. However, dependency between projects is more often the rule than the exception. For this reason, a single program often constitutes an important component of a technology strategy. If that program is a major success, the option chosen will influence, through its implementation, the choice of other technological options.

The DANTE Model helps to construct good configurations, contributes to the identification of the network of technologies and equipment, and evaluates the options. Through the use of the strategic criteria the DANTE Model provides an impetus for managers to integrate the technology program into the business and corporate strategy. This is not an "operational" improvement, but helps to create a competitive advantage which in fact is crucial to business strategy.

In contrast, if an important technology program is a failure, a deterioration of the organization's position compared with the competition often follows. Such a failure may endanger the vary existence of the organization. In either case the linkage between the success or failure of the technology program to corporate strategy is clear. Hopefully, the technology program will improve the company's position relative to the strengths of its competitors (Andrews 1980).

4.2 Contributions to Strategic Planning

A major activity of the strategic planning process is to allocate resources to program activities which can achieve a set of business goals in a competitive environment. This requires action plans and the integration of the annual planning and budgeting activities. In the case study, the organization had for many years separately prepared the plans and budgets. The application of the DANTE Model has required a comparison between the actual results and the plan, so that the resources can be reallocated to better achieve the objectives of the program. The different spending curves were estimated to balance overall resource usage. The triangle resource (i.e., technological, human and financial) tool proposed for the ALADIN Method may be applied with the DANTE Model (Danila 1988a). The multi-dimensional method ALADIN works with checklists, matrices, multi-criteria, consensus and graphical aspects of the problem.

4.3 Institutional Learning

Institutional learning is the process whereby management teams change the mental models of their organization, markets and competitors (DeGeus 1988). The DANTE Model helps managers to understand what is happening in their environment by measuring and tracking the various criteria and then taking appropriate action. The application of the DANTE Model helps to accelerate the decision-making process and fosters learning, providing an important competitive advantage. The Model creates new levels of communication among team members and enables them to express the knowledge they have acquired. Some users have stated that the model acted as a catalyst because it helped them reassess their business and the decisions they face. The application of the steps of the DANTE Model led managers to challenge established approaches and was the beginning of the institutional learning process.

4.4 Creativity in the Evaluation Process

The DANTE Model explicitly requires managers and program evaluators to expand their thinking and exercise creativity in identifying new options and configurations. These same configurations are then evaluated in the later steps of the model. Thus, this approach directly links the analytical evaluation process with the creative and partly intuitive process of breaking with the status quo and proposing new options and configurations. It is very important that the evaluators and program team members carefully emphasize the utilization of intuitive methods during the first step of the model.

The importance of the first step in identifying options stresses the need for evaluators and managers to be innovators as well as analysts. Analytical tools are used to stimulate the creative process and test the resulting ideas to help ensure that the technology program selected can achieve the strategic objectives and be implemented.

4.5 Limits of the DANTE Model

The DANTE Model has three principal limitations. The first concerns the use of intuitive approaches and tools throughout the steps of the model (see Table 3). Intuitive techniques and approaches are difficult to assess and even more difficult to teach or develop within individuals. Some creativity approaches can be taught but a corporate culture must be cultivated in these approaches are to be fully applied.

A second limitation is more generic to evaluation processes which require a variety of activities by different individuals. Specifically, the need for timely evaluations is crucial if managers are to make reasoned evaluations of all information relating to specific technology configurations. Delays can cause incomplete evaluation of some configurations or overall delays in reaching implementation.

Finally, the flexibility of considering a variety of criteria is at once a strength and a weakness. It is the latter because decision-makers do not always wish to divulge all of their evaluation criteria, enabling them to exercise additional control over the final decision. Therefore, some risks exist as a result of insisting on the choice of the criteria themselves, their weights, and the parameters for the concordance and discordance tests.

5. CONCLUSIONS

The DANTE Model appears to be well-suited to the selection and evaluation of technology programs which consists of dependent options. It has been applied successfully to the problem of evaluating complex configurations of technologies and equipment. The model combines analytical and intuitive thinking throughout its seven steps.

Thomas Edison's maxim concerning invention being "1% inspiration and 99% perspiration" seems appropriate for the DANTE Model. The "inspiration" concerns the identification of options. The "perspiration" corresponds to the collection of data, and the phases of evaluation leading to the selection of the best option and its implementation.

The process of implementing a successful technology strategy can be likened to the making of a great motion picture. Movie-making calls for technical mastery of the various activities required, but the director's work, the "mise en scene," does not have to be observed. Great works originate in insights which are often intuitive rather than rational. From this point of view, the DANTE Model can be seen as not another evaluation technique but a method for performing a strategic evaluation. The suggested approach for strategically managing technological resources is to view technology options as a network of technologies and equipment. In spite of some limitations, the DANTE Model can help initiate an institutional, creative, strategic-oriented learning process and build a successful technology strategy.

6. REFERENCES

Andrews, Kenneth R. (1980). **The Concept of Corporate Strategy.** New York: Dow Jones-Irwin.

Archibald, Russell D. (1976). **Managing High-Technology Programs and Products.** New York: John Wiley and Sons.

Danila, Nicolas V. (1989). "Support Graph for the Managerial of High-Technology Projects," **Technology Analysis and Strategic Management, 1** (3), 213-284.

Danila, Nicolas V. (1988a). "Multicriteria Methods for R & D

Management, **Joint International Conference: EURO IX - TIMS XXVIII**, July 6-8, Paris, France.

Danila, Nicolas V. (1988b). "ALADIN, A Tool for Strategic Planning," **Eighth International Symposium on Forecasting**, June 12-15, Amsterdam, The Netherlands.

Danila, Nicolas V. (1985). "Evaluating Pharmaceutical R & D," **Managing High-Technology: An Interdisciplinary Perspective**, B. W. Mar, W. T. Newell, and B. O. Saxberg (eds.). North Holland: Elsevier Science Publishers.

Danila, Nicolas V. (1983). **Strategies Technologiques: Methodes d'Evaluation et de Selection des Projets de Recherche.** FNEGE.

Danila, Nicolas V. (1979). "Methodologie d'Aide a la Decision dans le Cas d'Indicateurs fort Dependants," **Paris IX Dauphine**, Unpublished Doctoral Disertation.

DeGeus, Arie P. (1988). "Planning as Learning," **Harvard Business Review**, March-April, 70-74.

Helmer, Olaf. (1966). **Social Technology.** New York: Basic Books.

Kerzner, Harold. (1984). **Project Management: A Systems Approach to Planning, Scheduling, and Controlling.** New York: Van Nostrand Neinhold Company.

Roy, Bernard. (1985). "Methodologie Multicritere d'Aide a la Decision," **Economica**, Collection Gestion, Paris.

THE ECONOMIC EVALUATION OF ADVANCED MANUFACTURING TECHNOLOGY

P. L. Primrose
Total Technology Department
University of Manchester Institute of Science and Technology
P.O. Box 88, Manchester M60 1QD U.K.

ABSTRACT

Companies that have installed Advanced Manufacturing Technology often find that their investment has not been viable. The reasons for this are examined and it is shown that the problem arises because of the way companies evaluate and cost the projects. The nature of the problem is explained and suggestions are made for the solution. The conclusion is drawn that there is nothing wrong with investment appraisal and costing principles, rather the underlying cause is the incorrect application of the principles.

1. INTRODUCTION

All the literature aimed at production engineers helps to reinforce their belief that Advanced Manufacturing Technology (AMT) is essential to ensure their company remains competitive. Typical of this message is the National Economic Development Office's report on AMT (1985) which suggests that the benefits of each area of AMT are so large that every company should install as much AMT as is technically possible.

At the same time, companies that have installed AMT are reporting a lack of financial success: A recent British Institute of Management survey (New and Meyers, 1986) of 239 companies suggested that nearly half were getting poor or negative returns from their investments. In the cases of Flexible Manufacturing Systems (FMS) and Robotics, the respective figures were two-thirds and three-quarters of companies reporting low or negative payoff.

While many authors, such as Kaplan (1984), have identified the fact that there is a problem, they have been unable to identify the basic underlying causes, the danger being that any attempt to cure the perceived symptoms will only make the situation worse.

The assumption normally made is that existing accountancy systems are somehow inadequate so that a radically new approach is needed, such as the development of the CAM-I Cost Management System in America or of Throughput Accounting in Britain.

Engineers advocating AMT investment are faced with two financial problems: first, how to convince accountants that the investment should be made and second, once the investment has been made, the subsequent discovery that the project appears to be a financial liability.

To understand the nature of the problems faced by engineers, it is necessary to start with the basic criteria that have to be fulfilled by an AMT project, namely:

1. The project must generate sufficient cash to ensure that the initial cost is repaid, all interest charges are paid, and sufficient profit is made to compensate for any element of risk.
2. The investment is correctly reflected in product selling prices.
3. Costs and savings are monitored to ensure that forecasts are achieved.
4. The investment is correctly reflected in the company Balance Sheet.

The cause of much of the perceived failure of AMT is that these criteria are not met. In such situations, the project will appear to be a financial failure, regardless of its true worth, or how essential it is to ensure the company's long-term viability. Therefore, it is necessary to examine these criteria in order to identify the nature of the problems faced by companies, and to show how they can be resolved.

2. INVESTMENT APPRAISAL

Engineers normally treat investment appraisal as a final pass/fail hurdle that has to be overcome. Having spent considerable time establishing an AMT project's specification, normally based on technical objectives, and having, in the process, become emotionally convinced that the investment is essential, there is a strong temptation to adopt the attitude of "how large a figure do I need to get it 'past' the accountants?".

In such a situation engineers often allow themselves to be persuaded by suppliers to accept unrealistic estimates of savings based on a typical or theoretical examples. However, investment appraisal has to be treated as an aid to management decision-making that can do much more than provide a simple yes/no verdict on a project. The objectives in using investment appraisal are:

1. To help select the most appropriate area of AMT for initial investment.
2. To help choose the correct specification and supplier.
3. To help establish measurable objectives and a timetable for implementation.
4. To quantify all costs and savings so the investment is correctly reflected in the company's costing system.
5. To ensure the investment will be profitable.

Companies have a wide choice of technology such as Computer Aided Design (CAD), Computer Numerical Controlled (CNC) machine tools, Material Requirement Planning (MRP), Manufacturing Resources Planning (MRPII), Flexible Manufacturing Systems (FMS) and Robotics. Although generally referred to as AMT, not all of these choices are directly concerned with manufacturing.

It follows that the first problem faced by a company is to identify real needs and select where initial resources should be concentrated, rather than just installing technology which is currently fashionable, or responding to pressure from individual managers who are only considering narrow departmental interests.

One outcome of the work done at the University of Manchester Institute of Science and Technology has been to prove that there is nothing wrong with establishing principles for investment appraisal, but that problems have resulted from incorrect application of those principles. The following three problem areas have been identified and will be described below:

1. Use of incorrect methodology.
2. The belief that conventional investment appraisal could not be used for AMT.
3. The use of alternative selection techniques.

2.1 Incorrect Methodology

As described by Primrose & Leonard (1987), companies have consistently understated the benefits of AMT, often by over 100%, by using incorrect methodology. The main errors are as follows:

1. The use of Payback rather than Discounted Cash Flow (DCF).
   The problems caused by using Payback for AMT projects is not just of theoretical importance since it can result in companies failing to invest in projects which are vital for its long term viability. Much of AMT, such as CAD and MRPII, can take one or two years to install before the full level of benefits are realized. However companies using Payback normally only select projects which give a one or two year payback period. Despite this, a survey of 100 major UK companies by Pike & Wolfe (1988) showed that a majority still used Payback.

2. The criteria used for project acceptance.
   Although much has been written about methods for calculating the cost of capital, companies often use irrational criteria. For example, a large firm stated that it required both two year Payback and 15% Internal Rate of Return (IRR). By doing this, a project with an expected life of ten years, which could generate over 50% IRR, may be rejected because the timing of initial cash flows does not provide a two year Payback.

3. The use of non-cashflow information, such as depreciation and unaltered overheads.
   A major contribution to this problem is that much of AMT is used for "cost reduction" and, as such, engineers use standard cost data from an Absorption Costing system in their justification because this information is used to measure their department's

performance. A typical statement made by engineers is "we decided not to replace that machine because it was costing £20 an hour to make the parts here and we can sub-contract them for £15 an hour". This argument ignores the fact that possibly £10/hour of in-house costs represent fixed overhead which still exist when parts are sub-contracted. The cause of such problems is not costing systems, which were established for product costing, but the use of standard costs by engineers in making decisions where variable costing should be used.

4. Incorrect valuation of inventory changes.
   Inventory reduction is portrayed as a major objective of AMT, but it has been found that the treatment of the subject in most accounting literature is incomplete (Primrose and Leonard, 1987). The value of inventory reduction is strongly influenced by the methods used to reduce inventory levels. There are also considerable tax advantages in inventory reduction which must be considered in the analysis.

5. Use of Spreadsheets.
   The complexity of evaluating AMT projects, where 20 to 30 factors may have to be included, means that when engineers have attempted to use spreadsheet-type programs the resulting evaluation is over-simplified. As a result, they inevitably understate profitability, or make mistakes, thus increasing the engineer/accountant conflict. As pointed out by Schofield (1987), the use of spreadsheets can lead to major errors and it is very easy to be misled by them.

2.2 Inability to Evaluate AMT

There is widespread belief that conventional accountancy cannot be used to evaluate many of the benefits of AMT which are thought to be intangible. As a result, several alternative approaches have been advocated. When AMT benefits are stated in general terms, such as "better quality products", "improved management control" and "increased flexibility of production", it seems impossible to quantify these in financial terms as part of an investment appraisal.

The most common view is that AMT investment has to be treated as a strategic decision. Such a view, like the Lorelei, has a seductive charm because, not only is it much easier to make such a decision, but, few managers would admit that their strategic decision, while being technically elegant, may be a financial disaster. Unfortunately, the history of AMT is full of such strategic disasters.

Primrose and Leonard (1987) have shown that the major financial benefits of AMT come, not from improving manufacturing efficiency, but from the effect it can have on making the total company more competitive. Because of this, projects must be considered in the context of overall company planning. Selecting and evaluating projects to suit a company's long term strategy, such as five or ten year plans, is very different from making a decision purely as an act of faith (or quite often an act of desperation), and then calling it a strategic decision.

An alternative to the act-of-faith approach requires making a check list of benefits and then give them a score of 0 to 10 depending on their perceived importance. While being slightly more organized than the act-of-faith approach, the decision is still purely subjective. An improvement on this is suggested by Kaplan (1986), and requires calculating how large the savings need to be and then consider whether the number is feasible. The inherent danger of this is that it uses the same philosophy as engineers who say "how big a number do I need to get it accepted?", and is based on trying to justify the engineer's decision rather than evaluate alternatives.

2.3 Quantifying Intangible Benefits

The perceived difficulty of evaluating AMT stems from the belief that some benefits are intangible and cannot be quantified. However, Primrose & Leonard (1987) have shown that EVERY benefit which can be identified can be re-defined in such a way that it can be quantified and included in a conventional investment appraisal.

No benefit should ever be excluded on the grounds that it is intangible. To do this, the benefits have to be dealt with in a two stage process:

1. Re-define into quantifiable terms, and
2. Calculate the magnitude of the value.

The problem has now changed from the inability to include benefits to that of the accuracy with which they can be measured. This is a completely different and manageable problem where techniques, such as sensitivity analysis, are available to help.

A simplified example of this re-defining process is the benefit of achieving "better quality products" which can be restated as:

1. Reduced Cost of Scrap;
2. Reduced Cost of Re-work;
3. Reduced warranty and service costs;
4. Reduced cost of lost production carried by scrap and rework;
5. Reduced inspection and quality control costs; and
6. Increased sales of better quality products.

When these items are quantified in actual projects, the value of the sales increase can be much greater than all the other factors combined. The example of Jaguar Cars shows the effect which an improved reputation for reliability has on sales.

In the case of "increased flexibility of production", previous efforts had concentrated on trying to develop a measure such as a percentage of flexibility. The work at UMIST was able to identify 25 different aspects of flexibility and to show that every one of these was part of the technical specification. What has to be quantified is not flexibility, but the benefits, such as the effect on sales, which the technical specification (including flexibility) can provide.

The ability to evaluate all the benefits of FMS has shown that investments in multi-machine systems can rarely be justified on the basis of production savings. The real potential of FMS is the effect which it can have on delivery performance and the consequent increase in sales.

2.4 Alternative Techniques

Conventional investment appraisal is based on calculating the value of costs and savings and then using these to calculate a DCF return. This return is then used as the criteria to assess profitability and for comparison between alternative investment projects. An alternative approach is used in some single product industries, such as electricity generation and supply, where there is only one major criterion - the unit cost of supplied electricity.

The technique used is to calculate this unit cost based on alternative generating methods. Investment is then made in the method which produces the lowest cost. Although this approach can be valid within narrow constraints, it has been shown by Thabit & Primrose (1986) that when it is applied in an industrial situation, such as generating electricity for a company's own use, the results can be completely misleading - leading to incorrect investment decisions.

A variation on this technique is described by Nixon (1986) as being used within IBM to justify automated assembly equipment. The unit product cost is calculated for both manual and automated assembly, and investment in automation is then made if the product cost from the second method is the lowest. While the use of this technique overcomes the problems of incorrect use of depreciation, there are still two fundamental problems inherent in such a methodology.

As stated earlier, a major benefit of AMT is the ability to make a company more competitive, thereby increasing sales. Thus a comparison of unit product costs, with and without AMT, must include the effect on overheads of increased sales volume. However, as pointed out by Kaplan (1984), the allocation of overheads, especially in a multi-product organization, can be highly inexact, to say nothing of being a highly complex and laborious process.

The investment decision will be made on the basis of which method provides the lowest product cost, but the differences between costs may be very small. For example the costs may be:

£999 With AMT, and £1,000 Without AMT

As the AMT cost is lower, investment would be made in AMT; but a 0.2% difference in costs could have reversed the decision. Unfortunately, inaccuracies of 10% are quite possible, especially if some of the plant is to be used for more than one product.

The second problem is that, even if the correct product costs could be calculated, they would be meaningless when trying to compare alternative investments which can affect different products in different ways. One of the basic reasons for using investment appraisal is to be able to compare alternative investment strategies (e.g. CNC or FMS or Robotics).

2.5 Correct Methodology

The reason why the concept of intangible benefits developed is that the benefits normally occur in a different department than that in which the investment is made. In addition, the relationship between cause and effect is indirect so that, unlike many direct savings, the value of savings has to be estimated rather than calculated.

In the work on investment appraisal at UMIST, it was found that once the problems had been identified, the underlying cause was the mutual difficulty in communication between engineers and accountants. The engineers were unable to explain how AMT could affect the company in financial terms, while the accountants were unable to explain how financial systems could be used to increase the effect of AMT projects.

To overcome the problem of communication, IVAN, an investment analysis microcomputer program, was developed for use by engineers (see listing in Section 7). This program is now being widely used and the user list is beginning to look like a Who's Who of British

Engineering. IVAN acts as a tool to overcome the communication problems by presenting projects in terms mutually understandable to engineers and accountants.

In order to identify the nature of the problem of intangible benefits, an extensive literature search was conducted. The aim was to discover all the benefits which previous authors had identified. Comprehensive lists of these benefits were compiled for each aspect of AMT and then redefined into quantifiable terms.

In addition to benefits, all potential areas of cost were also identified. Lists were then produced of all costs and benefits of each aspect of AMT. For example, the list contains 63 factors for CAD/CAM, 43 for Robotics, 90 for MRP/MRPII and 85 for FMS. These lists were then combined and incorporated into IVAN.

In doing this, care had to be taken not only to ensure that all potential costs and benefits were included, but also to avoid any duplication. Several factors can lead to the same type of saving; for example, both improved product specification and improved delivery performance may result in increased sales. At the same time, a single factor can lead to several types of savings. For example, shorter lead times may both reduce inventory and increase sales. To try to avoid duplication, the lists of benefits used to create IVAN was divided up into the types of savings, rather than by the causes.

Data input to IVAN avoids the use of non-cashflow information, such as depreciation and unaltered overheads, while calculation of factors such as taxation and DCF return are performed automatically. Since IVAN is designed solely for investment appraisal, users cannot alter its operation. This helps to overcome the problems associated with the use of spreadsheets mentioned previously. IVAN can be used by managers without financial training. However, since it adheres strictly to accountancy principles, the results can be checked by, and be acceptable to, accountants and financial managers. Although IVAN can be used to evaluate ANY manufacturing investment project, ALL factors affected by the investment must be included.

The use of IVAN has confirmed that there is nothing wrong with investment appraisal principles. IVAN's application in a wide range of projects has also confirmed that all potential benefits can be quantified and has shown that the benefits previously excluded as intangible are often the most important ones. A brief case study describing the application of IVAN to an investment in MRPII can be found in the Appendix.

Because engineers can now evaluate projects (using estimated data) at the start of their investments, they no longer waste time on non-viable projects, and thus subsequent conflict is avoided. As a result, the most important factors can be identified at the outset so that work can be concentrated on selecting technology for the correct objectives.

Although a large number of potential benefits may be identified initially, there is normally a marked pareto effect. That is, only a very small number of factors, possibly only one or two, have a significant effect on the result. By concentrating on these, the accuracy of estimates can be improved, thereby increasing the accuracy of the evaluation.

3. PRODUCT COSTING

One of the most commonly reported problems associated with AMT is that companies have installed major projects only to find that it has resulted in an increase in product cost. The conclusion is then reached that either AMT is a failure or the costing system is incorrect. Primrose (1988) has now shown that in fact neither conclusion may be correct.

If an investment is made as an act of faith, without a financial appraisal, it cannot be correctly recorded in the product costing system. Even without an appraisal, all capital, installation and running costs will be accurately recorded and attributed to the project because the bills will have to be paid.

Because savings have not been forecast and quantified, they cannot be attributed to the project. Therefore, the costing system must show that the investment has increased product costs. If savings had been quantified as greater than costs, there would be no need to invest as an act of faith.

As previously stated, the nature of intangible benefits is such that they will occur elsewhere in the company than the department where the investment is made. In addition, there is likely to be a time lag of one to two years between making an investment and the full level of savings being achieved. This means that when the savings do occur, most of them will just appear as a general operating variance and not be attributed to the project.

For all of the reasons it is clear that if an investment is made as an act of faith, without an evaluation, the costing system MUST show that it has increased product costs.

An additional problem with costing systems is seen in companies that have gradually changed from being labor intensive to capital intensive, without changing the basis of their costing system from labor/hour rate to machine/hour rate. The problem is highlighted with FMS which has the ability to run independently of the human operators, and where manning levels can vary between shifts. As a consequence the relationship between labor costs and machine/hour costs becomes meaningless.

When the reasons for the apparent failure of costing systems were investigated, it was found that the problems do not lie in any failure of costing principles. Rather the failure results from the incorrect application of the principles by companies that have not updated their systems to match the way the company has changed. Unfortunately, while there is still a belief that "there is something wrong with costing", companies will refrain from updating their procedures as they will be unsure that what they are going to do will be the right thing.

3.1 Life Cycle Costing

In the past, companies bought a plant either for general use or, as in the case of transfer lines, to produce a specific product. When the intention was for a product to remain in production for many years, it was assumed that the plant bought specifically for the product would be scrapped when production ceased. In such cases product life and physical plant life would be the same. Therefore, the decision to invest in the plant would be an integral part of the decision to invest in the product with capital costs being included in product costs.

When the plant is bought for general use and used for several different products, it would be kept until the end of its physical life and this would have no relationship with product life.

Due to pressure on companies to keep introducing new products, expected product life is decreasing, resulting in an increasing trend away from dedicated automation towards flexible plant which can be quickly converted to introduce new products. The result is that the relationship between product and plant life is changing. Companies in high volume markets must be aware of this and change the way they allocate capital costs to products as physical plant life may now represent several product life cycles.

4. RECORDING COSTS AND SAVINGS

The use of a post-audit to compare planned costs and savings with actual results is essential but it must be treated as an aid to improving future decisions, and not as a way to apportion blame for any failure. The use of IVAN to post-audit projects has identified a number of problems which have made some projects less successful than predicted. The main problems are:

1. Insufficient allowance for start-up time.
   Although it may only take a few months to install projects such as FMS, CAD and MRPII, it can take a long time, often two or three years, before full savings are being achieved. The problem is particularly acute in companies that use Payback as an evaluation technique. The use of Payback encourages managers to be overly optimistic about start-up times.
2. Insufficient allowance for running costs.
   With computer systems, such as CAD and MRPII, the running costs over the project's life can exceed initial capital costs; however, engineers often only justify the initial cost.
3. Use of unrealistic savings.
   An example is that the literature relating to CAD emphasizes increased draftsmen productivity. A ratio of 3:1 between drawing board and CAD is often quoted, and consequently used by companies in their evaluations. However, experience of post-

audits is now showing that a ratio of 1.5:1 is more realistic.

While there can be an element of over-optimism from engineers trying to get projects accepted, a major cause of the above problems lies in the fact that engineers are often planning the introduction of new technology about which they may have little or no operating experience. The use of post-audits can identify failures such as these so that advice can be given to industry to help prevent the problems from recurring.

5. EFFECT OF INVENTORY REDUCTION

Inventory reduction is often cited as a major benefit of AMT, especially in the current spate of literature about Just-in-Time (JIT); but such reductions can have serious implications for a company's Balance Sheet. Because of the way inventory changes are recorded in the double entry bookkeeping system used by most companies, the material, labor and variable overhead elements of inventory reduction are shown in the Balance Sheet as a change in assets, not a reduction in operating expenses.

The reduction in the fixed overhead element of the inventory value, which may comprise 50% of total book value, will be reported as a reduction in profit. The result is a paradox whereby inventory reduction, which is always portrayed as "a good thing", can result in a reduction in reported profits while at the same time generating a positive cash flow due to the reduced tax liability.

The effect of inventory changes on a company's accounts can be complex, with revaluation of inventory often being used in "creative accounting" to produce short-term cosmetic changes. It is therefore essential that engineers considering any project where inventory reduction is a major element discuss the full implications with the company's accountants.

6. CONCLUSIONS

When AMT projects are selected to solve genuine company problems, and they are correctly evaluated and costed, the investment can be very attractive. If evaluation and costing are not done correctly, it is likely that projects will be selected for the wrong objectives with a sub-optimal specification being chosen. Even if such projects happen to be financially viable, the costing system is likely to report that they are not viable.

Because of the complexity of AMT, and the way it is reflected in a company's financial systems, there is a danger that the true nature of the problems encountered in their evaluation and costing is not correctly identified. Corrective action is then aimed at the symptoms rather than the basic cause. Companies need to be able to identify the causes of any apparent failure of AMT so that the correct action can be taken.

There is a need to educate engineers in the nature of cost systems and accountants in the financial effects of AMT. The present mutual lack of understanding between engineers and accountants has led to the universal belief that accountancy principles are outdated when applied to AMT; happily, such a belief is quite wrong. The problems do not lie in any failure of the basic principles but in the outdated procedures used by companies.

7. REFERENCES

Kaplan, R. S. (1984). "Yesterday's Accounting Undermines Production", **Harvard Business Review**, July/August, 95-101.

Kaplan, R. S. (1986). "Must CIM Be Justified By Faith Alone?" **Harvard Business Review**, March/April, 87-95.

IVAN - Investment Analysis Computer Program. Applied Technology Ltd, Carrington Business Par, Urmston. Manchester, UK.

National Economic Development Office. (1985). "Advanced Manufacturing Technology," London.

New, C. C. and A. Myers. (1986). "Managing Manufacturing Operations," British Institute of Management.

Nixon, W. (1986). "MEI - Manufacturing Early Involvement," **Proceedings of I.Prod.E Seminar on "Justifying investment in**

Automated Assembly", February.

Pike, R. H. and M. Wolfe. (1988). "Capital Budgeting For The 1990's," **Institute of Cost and Management Accountants.**

Primrose, P. L. (1988). "The Effect of AMT Investment on Costing Systems," **Journal of Cost Management for the Manufacturing Industry, 2** (2), 27-30.

Primrose, P. L. and R. Leonard. (1987). "Performing Investment Appraisals For AMT," **Journal of Cost Management for the Manufacturing Industry, 1** (2), 34-42.

Schofield, J. (1987). "Beware of Spreadsheets," **Management Today**, February, 39-40.

Thabit, S. S. and P. L. Primrose. (1986). "Conditions Under Which Wind Turbines Can Be Financially Viable For Private Power Generation In Industry," **Proceedings of I.Mech.E, 200A** (2), 109-115.

8. APPENDIX

Case Study - Investment in MRPII

The company manufactures process equipment for the food industry and has an annual turnover of £13,000,000. Sales are dominated by a small number of large contracts, each worth several £million. As a result, the work load on the factory suffers from extreme peaks and troughs, with extensive sub-contracting being used to deal with the peaks.

In recent years, the company has installed several micro-computer based systems for Stock Control and Accounting applications, and a mini-computer based CAD system. All the existing systems, except CAD, are very limited in the functions they can perform, so that the intention is to replace them all with a single MRPII system. To install the new MRPII system, changes will be required in each department directly affected, as well as in some aspects of overall company operations. Because of this, it is assumed that it will take 18 months before the company starts to obtain the financial benefits of the investment. By concentrating on the major savings, these are achievable within this time frame, although it may take three years to reach full installation.

Initial Costs

Hardware (capital)	£123,800	
Software (capital)	£150,700	
Computer Room (capital)	£10,000	
Installation & Wiring	£10,000	year 0
Training	£2,300	year 0
Training	£6,000	year 1
Customizing Software	£4,000	year 0
Data entry & Start-up	£10,000	year 1
Data entry & Start-up	£5,000	year 2

Annual Running Costs (from year 1)

System Management	£20,000
Maintenance	£33,000

It is assumed that the running costs (consumables, electricity, etc.) will be paid for by replacing the existing systems.

Savings

The anticipated major savings are as follows:

1. The current time from the receipt of an order to documentation being issued onto the shop floor represents nearly 50% of the total delivery time. By speeding up the pre-manufacturing stages, the manufacturing lead time can be increased by at least 50%. This, combined with improved shop floor scheduling will increase machine and labor utilization, resulting in a reduction in sub-contracting. Reducing the present £1,200,000 annual cost of sub-contracting by 10%, (minus the variable cost of producing the work in-house), gives a saving of £42,000 in year 2 and £84,000 per year from year 3.

2. The present system for estimating prices for quotations is very imprecise. Because the company will always tend to receive the orders for which they have underquoted, many orders are of marginal profitability. By improving the data

available from the Costing System and improving the estimating and quotation procedures, it is estimated that the profitability of orders can be increased by at least 1%, giving a saving of £65,000 in year 2 and £130,000 per year from year 3.

3. The new system will ensure that all changes requested by customers, after they have placed an order, are correctly costed and charged to the customer. This will increase the profitability of orders by 0.5%, giving a saving of £32,500 in year 2 and £65,000 per year from year 3.

4. Transcription errors will be eliminated and the accuracy of the Bill of Materials will be improved. This will eliminate the ordering of unwanted components, while at the same time avoiding the present situation where shortages are found in Assembly Department because some components had not been ordered. These improvements will yield a saving in production costs of £18,000 in year 2 and £35,000 per year from year 3.

5. Improved capacity planning will allow the company to quote for additional orders which can be used to fill in some of the troughs in the workload. It is estimated that sales can be increased by 1%, giving a contribution to overhead recovery of £26,000 in year 2 and £52,000 per year from year 3.

6. Avoiding the manual transfer and re-entry of data between systems will provide a labor saving, but this may be offset by the additional stores and shopfloor clerical labor needed to improve the accuracy of data. As a result, no allowance is made for labor saving.

7. The new system will allow departments, such as Purchasing and Quality Control, to become more efficient and thereby reduce costs. However, no allowance has been made for these savings.

8. The increase in manufacturing lead time will increase the amount of Work in Progress, but it is thought that this will be offset by a reduction in raw material stocks and improvements in scheduling Purchase orders. Therefore, no

allowance is made for any changes in inventory.

Assuming a ten year project life, 35% tax rate and 12% cost of capital, Table 1A shows the cash flows used in the evaluation. These give an Internal Rate of Return (IRR) of 43.3% and a Net Present Value (NPV) of + £649,365.

The evaluation was performed using the IVAN software package. In order to find out how sensitive the results are to variations in the estimated savings, some of the factors were altered in turn. If the reduction in sub-contract was 5% rather than 10% the IRR would be 38.4%, a quite acceptable return. If the increase of profitability from improved estimating was 2% rather than 1% the IRR would be 56.6%. If the increase in sales was 5% rather than 1% the IRR would be 63.7%. These last two results show great upside sensitivity to small increases in profitability and sales.

Table 1A. Annual Cash Flows

	Year 0	Year 1	Year 2	Year 3	Year 4 on
Capital cost	-284,500	-	-	-	-
Capital tax	-	+24,894	+18,670	+14,003	Reducing
Installation	-14,000	-	-	-	-
Training	-2,300	-6,000	-	-	-
Start-up	-	-10,000	-5,000	-	-
Running costs	-	-53,000	-53,000	-53,000	-53,000
Less sub-contract	-	-	+42,000	+84,000	+84,000
Better estimates	-	-	+65,000	+130,000	+130,000
Customer's changes	-	-	+32,500	+65,000	+65,000
Wrong ordering	-	-	+18,000	+35,000	+35,000
Extra sales	-	-	+26,000	+52,000	+52,000
Revenue tax	-	+5,705	+24,150	-43,925	-109,550
Net cash flows	-300,800	-38,401	+168,320	+283,078	Reducing

PERFORMANCE MEASUREMENT AND PRODUCT COSTING IN THE AMT ENVIRONMENT: A LITERATURE REVIEW

James Borden
Dept. of Accountancy
College of Commerce and Finance
Villanova University
Villanova, PA 19085

ABSTRACT

U.S. manufacturers are making significant changes in the way they do business. Advanced Manufacturing Technology (AMT) enables firms to become more competitive in the international marketplace. However, many firms have found that their management accounting systems are woefully out of date. Such systems should be designed to provide relevant information concerning product costs, process control and performance evaluation. Obsolete management accounting systems (MAS), however, are not able to capture the realities of operating in an AMT environment. As a result, managers are faced with either dealing with irrelevant information for product costing and performance evaluation or developing new management accounting systems that provide relevant information.

The purpose of this paper is to explain why many of today's MAS are obsolete and to look at the many suggestions offered by leading academics and practitioners for improving the current state of management accounting systems for performance evaluation and product costing in an AMT environment. Ideally, the paper will help managers to recognize problems with their management accounting system as well as offer ways to help solve those problems. It is believed that improved management accounting systems will encourage more firms to invest in AMT and to realize the benefits these investments offer.

1. INTRODUCTION

Advanced Manufacturing Technology (AMT) offers the promise of enabling U.S. firms to achieve manufacturing capabilities so that they may succeed in a competitive environment. The inherent potential of such technologies may be lost, however, if a company's accounting system does not properly reflect the costs and benefits of operating in such an environment. AMT has created the need for a new set of performance measures, that reflect the financial and nonfinancial goals that have been set for management to achieve. AMT has also placed a premium on accurate product costing--the attempt to get a measure of "true" unit product costs. It has traditionally been the role of the company's management accounting system (MAS) to develop and monitor appropriate performance measures, as well to design the company's cost accounting system.

The decision to invest in AMT is most often not accompanied by an investment in the firm's accounting system. The likely outcome of such a situation is that performance measures become misleading and reported product costs begin to drift away from "true" product costs. The purpose of this paper is to summarize the role of accounting systems in an AMT environment. As such, it will look at how and why cost accounting systems were originally developed and why the majority of them have become obsolete. The paper will then look at the many suggestions that have been made, by both academics and practitioners, as to what performance measures are necessary in an AMT environment and how a company's cost accounting system can more accurately measure product costs.

Books and articles describing the problems with U.S. manufacturing and the relative superiority of Japanese manufacturing performance over U.S. manufacturing performance started to appear in the late 1970s and early 1980s. These early writings (Hayes and Abernathy, 1980; Hayes, 1981; Takeuchi, 1981; Wheelwright, 1981; and Schonberger, 1982) provided excellent insight into the necessity of U.S. manufacturers adopting new manufacturing technologies and philosophies. One could go back even further to the work of Skinner (1969 and 1974), in which the strategic role of manufacturing in competitive environments was discussed.

The availability of AMT has forced American companies to re-examine the role of manufacturing in their overall strategy. In response to the changing environment, American manufacturers have begun to adapt. Howell and Soucy (1987a) list six major trends evidenced by leading U.S. manufacturers: (1) higher quality, (2) lower inventory, (3) flexible flow lines, (4) automation, (5) product line organization, and (6) effective use of information.

However, the books and papers written in the late 1970s and early 1980s did not directly discuss the role of a company's accounting system in helping U.S. companies improve their manufacturing performance. There are two possible reasons for this disregard of accounting systems. First of all, as noted earlier, a majority of U.S. companies' accounting systems were, and still are, providing incorrect information to managers. Although operating managers may have been aware of these weaknesses in the early 1980s, they simply chose to ignore accounting data. Secondly, a company's accounting system was thought to have little impact, if any, on the company's strategic goals. Accountants merely played a support role, primarily geared to generating periodic financial statements.

A company's management accounting system needs to change in order to reflect the new manufacturing environment. However, it has become evident (Howell, et al., 1987) that many companies are still relying on obsolete and misleading management accounting systems, even in an AMT environment. Not only does this place accountants in danger of becoming obsolete as managers look elsewhere for useful performance measures, it also hinders companies from either making the initial investment in AMT or realizing the full benefits from investments in AMT.

2. THE EVOLUTION OF THE MANAGEMENT ACCOUNTING SYSTEM

To understand why current MAS do not provide the appropriate information needed in an AMT environment, it is helpful to take a brief look at the evolution of management accounting. Johnson

(1987) traces the beginning of management accounting to the early 1800s. The move from external market transactions to internal transactions during the Industrial Revolution (for example, having your own labor force as opposed to contract labor) provided the catalyst for cost management systems. Such systems were considered necessary if the relative efficiency of internal transactions were to be compared to market transactions. The information provided by these systems was tailored to meet management's needs.

With the advent of scientific management in the late 1800s and early 1900s, the next major force in MAS development was the focus on labor and material standards to control costs. Again, the data gathered was for the use of management in decision making, operations planning and controlling. However, as businesses became more complex, the burden of providing relevant information became restrictive.

At this same time, with the introduction of capital markets, external auditors came onto the scene. The auditors' methods of valuing inventory using objective criteria provided a lower cost alternative to costing products than techniques proposed by such people as Church (1917). Johnson does not believe that the demise of cost management resulted directly from the development of audited financial statements, but he does think that the requirement of such statements gave financial accounting a higher preference. Thus, the second- class status of cost management, combined with restrictive data processing costs, may have led to the demise of MAS.

Shortly after the initial wave of exposing the inferiority of U.S. manufacturing, the first major critique of U.S. accounting systems appeared. Kaplan (1983) noted that a company's survival may depend on the MAS responding to the changing manufacturing environment. AMT is **not** responsible for the breakdown of a company's management accounting system, but, rather, for bringing the limitations of current management accounting practices to the forefront.

Kaplan (1983) believes that accounting researchers need to develop measures of manufacturing performance that assess the key factors that affect a company's profitability in today's rapidly changing marketplace. New accounting systems that work in conjunction with a company's manufacturing policy, and not in opposition to the new production environment, must be devised. Improved measures of quality, inventory performance, productivity, flexibility and innovation will be required so that managers can focus on achieving long-term success and not be burdened by the current emphasis on short-term profitability. Kaplan's early work provided a starting point for academic researchers and others, such as accountants and production managers, to attack the problem of obsolete and misleading management accounting systems.

Miller and Vollman (1985) also looked into the problems of using traditional accounting methods in conjunction with advanced technologies. In particular, they examined the importance of managing factory overhead costs, which are becoming a greater percentage of production costs as companies adopt AMT. The authors refer to the costs incurred by off-line transactions as the "hidden factory." They propose that there is a need to move away from volume-related measures as allocation bases toward more of an activity- or transaction-based costing system.

Looking at the state of today's MAS, Johnson and Kaplan (1987) have classified MAS' shortcomings into three categories. First of all, current management accounting systems do not provide relevant information to help operating managers reduce costs and improve productivity. Secondly, these systems do not accurately measure the costs associated with manufacturing, marketing and distributing each of a firm's individual products--information critical to a manufacturing firm's survival. Many companies rely on the cost data produced for the external financial statements, which usually do not reflect the resources demanded by each product. Finally, Johnson and Kaplan believe that the emphasis on periodic financial statements forces managers to take a short-run view of profitability, at the expense of long-run viability.

The next section of this paper looks at the many suggestions that have been offered to enable manufacturers to correct these deficiencies. The focus is primarily on the first two, namely misleading performance measures and inaccurate product costing.

The issue of long-term views of profitability versus short-term profitability will also be discussed.

3. MANAGEMENT ACCOUNTING SYSTEMS IN AN AMT ENVIRONMENT

The following literature review is organized into two major categories: new and improved performance evaluation measures required by AMT and specific changes needed in a company's product costing system to capitalize on the benefits of AMT. While it is sometimes difficult to separate cost accounting issues from performance evaluation issues, enough appears to be written about each particular topic so as to warrant the attempt at such a division.

3.1 Performance Evaluation

Howell, et.al. (1987) note that if a company's short- and long-term performance measurements are not consistent with the ultimate objectives of management, then the likelihood of success is remote. A company needs to adopt measures that reflect the reality of operating in an AMT setting, and that are consistent with their long-run strategy.

Kaplan (1983 and 1984) has suggested that the following measures will prove useful in the new manufacturing environments:

Quality

1. number of defects at each stage of the manufacturing process
2. the frequency of machine breakdowns
3. percentage of completed goods that required no rework
4. the number and frequency of defects reported by customers

Inventory Performance

1. overall uncertainty in the production process by better scheduling of deliveries and production

2. the impact on inventory costs of having more flexible or more reliable production scheduling

3. cost savings from changing the parameters of the process itself, by attempting to reduce overall uncertainty

4. average batch sizes

5. the quantity of work-in-process (WIP) and raw materials at any time

Productivity

1. ratio of outputs produced to the physical inputs consumed

2. number of units produced

3. labor hours used

4. materials processed

5. energy consumed

6. capital employed

Flexibility/Innovation

1. ability to introduce new products

2. ability to respond quickly to customer requests for changing product characteristics

3. ability to deliver new products at high quality levels and predictable delivery schedules

As will be seen in the reviews that follow, many authors are suggesting the same basic measures proposed by Kaplan. Thus, controversy does not surround what should be measured in an AMT environment. The major problem lies with the ability of the management accountant to obtain the requisite performance measures and to then have management focus their efforts on the appropriate measures.

Howell, et al. (1987) take a slightly different approach than Kaplan in discussing relevant performance measures. Based on a survey as to the desired changes in performance measurement systems, the authors assigned the responses to one of the following categories: (1) the reporting process, (2) long-term orientation, and (3) content. The specific changes in each category are as follows:

Reporting Process

1. emphasize variance analysis
2. emphasize responsibility of individual managers
3. emphasize exception reporting
4. simplify measurement system and focus on key results

Long-Term Orientation

1. emphasize longer-term financial returns

Content

1. emphasize productivity measurements
2. measure cost of carrying inventory
3. measure cost of quality variances
4. measure manufacturing capacity utilization
5. introduce non-financial operating measures

Howell and Soucy (1987c) describe some of the non-financial operating measures that will be necessary in the new manufacturing environment. Under the major categories of (1) quality, (2) inventory, (3) material/scrap, (4) equipment/ maintenance, and (5) delivery throughput, the authors provide both a listing of non-financial, as well as financial, performance measures. For example, within equipment/ maintenance, the authors suggest four performance measures: (1) equipment capacity/utilization; (2) availability/downtime; (3) machine maintenance; and (4) equipment experience.

Bennett, et al. (1987) looked at the issue of performance evaluation by examining the problems that appeared to accompany the adoption of AMT. Bennett looked at four specific types of factory automation: numerical control (NC) machines, computer-aided design/computer aided-manufacturing (CADCAM), flexible manufacturing systems (FMS) and material handling systems.

Bennett described the problems found with each technology, as well as possible solutions to those problems.

First of all, the authors found it difficult to quantify the benefits of NC machines and to separate the cash flows from different machines. Also, it was difficult to evaluate machine utilization using statistics designed to evaluate labor. Finally, the researchers found that current evaluation methods focused on local goals, such as efficiency, rather than global goals, such as productivity. Bennett, et al., feel that new productivity measures that enable management to focus on the bigger picture while taking a longer-run view of profitability need to be developed.

Some new, quantifiable measures have been developed for evaluating CADCAM manufacturing activities. The authors found new standard cost figures being used that reflected improvements such as shorter set-up times, reduced material usage and reduced manufacturing time. However, benefits such as fewer defects and less spoilage, improved quality, more versatile tools and fixtures and development of automated quality control were not being measured, despite the fact that management was probably aware of these benefits.

For CAD, even less of a movement to develop relevant performance measures exists. Bennett, et.al., have suggested the following as possible criteria to evaluate CAD:

1. number of drawings produced
2. number of designs developed
3. number of proposals developed for marketing
4. time required to develop designs
5. number of NC programs developed
6. frequency of engineering change orders
7. time required to perform design analysis

In looking at flexible manufacturing systems (FMS), the authors proposed measurements such as:

1. machine and system utilization
2. productivity of the FMS
3. actual vs. planned throughput time per unit of product
4. manufacturing flexibility
5. quality, including percentage of defects and rework percentages
6. levels of work-in-process, raw materials, and finished goods inventory

The authors also note that, in order to ensure maximum FMS utilization, a diversified mix of products must be run on the FMS.

Finally, for performance measurement of automated storage/retrieval systems (AR/RS), a number of operating statistics are recommended, including the measurement of system operating time, computer operating time, computer picks per worker hour and an acceptable inventory accuracy percentage. In summary, Bennett et al. feel that accountants need to work more closely with manufacturing and engineering to develop new performance measures, which should be in place when the automation becomes a reality.

In another major look at performance measurement in AMT environments, McNair, et al., (1988) point out that product profitability should be the main focus. According to the authors, the present emphasis of a company's MAS on cost measurement will be replaced by a total performance measurement system that monitors flexibility, dependability, quality and cost. In keeping with their theme of a total performance measurement system, the authors have identified a wide array of performance measurement areas, such as:

1. design for manufacturability
2. zero defects
3. minimization of raw in-process inventory
4. zero lead time
5. minimization of process time

6. optimization of production
7. production linearity
8. zero set-up time
9. zero finished goods inventory
10. management cost structure
11. minimization of total life cycle cost

Within each of these areas, McNair et al. have identified one or more key characteristics. For example, in the design for manufacturability area, the authors list the following key factors:

1. quantity and quality of engineering change
2. test results
3. parts standardization
4. engineering cycle time
5. product complexity

In closing, it should be noted that the last three studies cited (Howell, et al., Bennett, et al., and McNair, et al.) were part of a major research project undertaken by the National Association of Accountants (NAA), along with groups such as Computer Aided Manufacturing-International (CAM-I) and Coopers and Lybrand. The studies represent findings from extensive questionnaires and field studies and represent an important contribution to the field of accounting. Another useful reference is the series by **Industry Week** about strategic manufacturing.

Although other contributions to performance evaluation literature exist, some representing partial findings during the course of research projects by the authors noted above, the studies cited provide a useful and comprehensive look at performance evaluation. The reference section provides a comprehensive listing of articles related to the topic of performance evaluation that the interested reader may want to pursue.

3.2 Product Costing

Johnson and Kaplan (1987) point out that a company's MAS must provide information on accurate product costs so that pricing decisions, new product introductions, the dropping of obsolete products and response to competitors' products can be done correctly and on time. In addition, the cost accounting system must provide data to facilitate cost control efforts and improve productivity.

Brimson (1986) acknowledges that one of the most important, yet least understood, roadblocks to implementing an automated factory is a company's cost accounting system. He believes that most systems do not provide the appropriate information needed to manage the factory of the future. A number of authors (Howell and Soucy, 1987b; Kaplan, 1988) have noted that cost accounting systems serve at least three distinct purposes: inventory valuation, product costing and process control. As such, it is not possible to use one set of numbers for all three purposes without adversely impacting at least one of the areas.

Seed (1984) was one of the earliest writers to examine how cost accounting systems must adapt as companies move towards automated production systems. He highlights five specific changes he believes are necessary in order for management to have more reliable product cost information. Among these suggestions are combining labor and overhead into one cost pool, use of more sophisticated cost allocation techniques and refocusing the control system on the areas that actually do control the costs of manufacturing.

Johannson (1985) discusses the notion of product life cycle accounting. The thrust of life cycle accounting is that the distribution of costs among categories changes over the life of the product, as does the potential profitability of a product. The most critical time for controlling a product's cost is in the development stage, even though the majority of the product's total life cycle costs have not yet been incurred. Accountants and managers must recognize the discrepancy between when costs are committed to--during the development stage--and when they are actually incurred--during the manufacturing and distribution phases. Although generally accepted accounting principles (GAAP) cannot be used for such a procedure, with shorter product life cycles and increased emphasis on cost control, it is expected that more attention will be given to product life cycle accounting with the adoption of AMT.

Howell and Soucy (1987) note that, in the new manufacturing environment, the use of multiple cost centers will make it possible to focus more clearly on the location and causes of operating control problems. In looking at product cost measurement, a firm must realize that all costs--manufacturing, engineering marketing, administration, etc.--are product costs. The authors see several changes in cost accounting systems. First of all, standard costs will be used for planning, but less for control purposes, as actual cost data will be monitored more closely. Secondly, more companies will use job-costing practices. Thirdly, full cost information will be used less, particularly for inventory valuation and financial information.

As companies adopt AMTs, the resulting shift in cost structure toward a greater percentage of fixed costs has led to research about the best way to trace these costs to the products. Two new terms that have crept into the cost accounting literature are activity-based costing and long-term variable costs.

3.3 Activity-Based Costing

Cooper (1988a,b and 1989a,b) has written extensively about activity-based costing (ABC). Cooper believes that ABC can remove or reduce the distortions caused by product diversity, such as volume or size, and improve the accuracy of reported product costs. The firms that are most likely to benefit from implementing an ABC system according to Cooper, would: (1) have a low measurement cost associated with obtaining the additional data required; (2) be in highly competitive markets; and (3) have a very diverse product mix. In discussing the number of cost drivers needed, Cooper asserts that the answer depends on a variety of factors such as:

1. desired accuracy of unit cost information
2. the degree of product diversity
3. relevant cost of different activities
4. degree of volume diversity
5. use of imperfectly correlated cost drivers

Finally, in terms of the type of drivers to select, Cooper suggests the following items be determined: (1) the cost of measuring the quantities associated with each driver; (2) the correlation of the selected driver to the actual consumption of the activity by the products; and (3) the behavior induced by use of the cost driver. Cooper vividly demonstrates the inaccuracies possible in a traditional costing system by presenting an example in which the unit product costs of a traditional system versus an ABC system differed 300 percent. Another advantage of the ABC approach is that it dovetails nicely with the two-stage procedure of cost allocation, which is discussed in a later section.

Turney (1989) notes that ABC allows product designers to understand the impact of different designs on cost and flexibility and to modify their designs accordingly. ABC supports the continuous improvement process by allowing management to gain new insights into activity performance by focusing attention on the sources of demand for activities and by permitting management to create a behavioral incentive to improve one or more aspects of manufacturing.

McNair, et al. (1988) note that, in an AMT environment, the distinction between direct and indirect costs will become clearer, resulting in fewer arbitrary allocations. The authors also envision two major changes taking place in an AMT environment. First of all, the majority (85%) of all costs for a new product/process will be committed prior to production. Secondly, product development costs will have to be recovered in a much shorter time because of shorter product life cycles.

McNair, et al., have also developed an extensive list of cost drivers. These cost drivers are activity measures, such as labor hours, that can be used to assign overhead costs to individual products. Although the complete listing of all the suggested cost drivers and performance measures is too lengthy to be included here, an example of product complexity measures includes:

1. number of components per finished product
2. number of manufacturing operations per finished product
3. number of tools required per finished product
4. life/cost of tooling per finished product

Johnson (1988), another leading advocate of ABC, describes how such a system involves tracing all costs, as best possible, to the activities that cause them. Under an activity-based costing system, a company will have many allocation bases for overhead costs, and not just a single allocation base, such as direct labor. While traditional costing systems assume that products cause costs to be incurred, activity-based costing assumes that products incur costs by the activities they require, and hence traces costs to products through these activities. The driving force for such a costing system is that, in the new manufacturing environments, economy of scope-- not economy of scale--is critical for success.

The importance of economy of scope and complexity led to the notion of long-term variable costs. Cooper and Kaplan (1987) observed that the most variable and most rapidly increasing costs were those traditionally classified as fixed. Long-term variable costs are those costs that vary with the number of transactions, such as machine set-ups, shipping orders, scheduling and so forth. The cost of such transactions is usually independent of the size of the transaction--it does not vary with the amount of inputs and outputs. It does, however, vary with the transaction itself. So, if a firm wants to introduce additional products, it will need larger support departments to handle the additional transactions. (See Goldhar and Jelinek (1983) for a thorough discussion of economy of scope).

Cooper (1987a, 1987b) has also been espousing the power of the two-stage procedure for allocating overhead costs to products. The first stage takes the cost of certain resources and combines them into appropriate and homogeneous cost pools. As an example, take two support departments, A and B. The costs incurred by these two support departments were $80,000 and $200,000, respectively. If 10 percent of support department A's effort is spent processing purchasing orders, then a cost pool for purchase order activity would be established, with 10 percent, or $8,000, of support department A's costs going into the pool. If support department B exerted 30 percent of its efforts on purchase orders, then 30 percent of its costs ($60,000) would go into the purchase order activity pool. The remaining costs in each support department would be treated in the same manner, setting up cost pools to reflect the activities of the support departments.

After this first stage is completed, the cost of all the company's resources are in separate cost pools. In the example started above, there is now $68,000 in the purchase order activity pool. The second stage then involves finding an appropriate cost driver to trace the costs to the products. The term cost driver indicates that the products drive the consumption of the resources, and should be charged for doing so. The cost drivers chosen reflect the activity that takes place in the various cost pools.

To continue the example, assume there were a total of 100 purchase orders for various products, and of that amount, 40 orders were for product X, 20 were for product Y and two orders each for 20 other products. Product X would be charged 40 percent of the costs that have accumulated in the purchase order activity pool, or $27,200, and product Y would be charged 20 percent, or $13,600. What the two-stage procedure has essentially done is allocate, or assign, the costs from the support departments to the individual products.

While the notion of the two-stage procedure is not new, it becomes critical for companies that are incurring more and more indirect costs as a result of AMT to use such a procedure correctly. The use of multiple cost pools and cost drivers should better enable management to trace their cost more accurately to the individual product lines. In summary, AMT will compel companies to use their cost accounting system for controlling critical costs, accounting for the cost of complexity, for providing useful information to control operations and for more accurately determining product costs.

3.3 Other Issues

In an AMT environment, accountants need to be concerned about other issues besides product costing and performance measurement. This section will briefly look at two other topics that need to be considered in the design of an useful accounting system.

The first topic is the issue of variance analysis. With the changing manufacturing environment, some variance measures will become obsolete or insignificant. A good example is the labor variances. With labor utilization and costs dropping rapidly, labor variances are not of great concern. On the other hand, other variances take on more importance. For example, material usage variance becomes significant since materials represents the majority of the final product cost.

Also, as Howell, et al. (1987) point out, executives feel more emphasis should be placed on variance analysis. Increased emphasis on variance analysis, together with the decreased importance of some currently used variances, should result in the development of new variance measures. As an example, Howell, et al., as noted earlier, suggest measuring a cost of quality variance.

An important role of accountants in developing these new variance measures is to match them to the previously mentioned performance measures, incorporating the performance measures into the annual budgeting process. The performance measures' inclusion would communicate to managers and employees what variables are important in an AMT environment, how these variables will be measured and what upper management's expectations are concerning those variables.

The second issue is the increased importance of focusing more on the long-term impact of management decision making. Howell, et al. (1987) argue that executives believe that long-term financial performance measures that are consistent with the long-term strategy of the firm need to be developed. The performance measures cited earlier are attempts to do just that--increase management's awareness of the long run.

Cooper and Kaplan (1987), as noted previously, introduced the notion of long-term variable costs. This term is appropriate since it gets managers away from thinking about traditional short-run periods of time and more toward analyzing the long-term impacts of their decisions. Turney (1989) also notes that activity-based costing provides information to aid management in making long-term, strategic decisions about choices such as product mix and sourcing.

While academics seem to be developing the appropriate tools for helping managers take a long-run view, another key aspect of management that also needs to be changed is executive compensation. Most salary and bonus plans are tied to short-term performance measures, which create strong incentives for management to be overly concerned about short-term performance. Thus, executive compensation plans that reward managers for making good long-run decisions also need to be developed. One such example has been the use of stock options redeemable some time in the future. While this plan seems to hold promise, the need for more long-term employee incentive contracts is obvious.

4. CONCLUSIONS

The adoption of AMT is having a profound influence on the way companies compete in today's international marketplace. Such technology allows for the production of higher quality goods, more diversified product offerings and greater flexibility, together with the promise of lower costs. This paper has reviewed the many suggestions that have been put forth in terms of developing relevant performance evaluation measures and improving product costing systems in order to take advantage of all that AMT has to offer.

Performance measures need to be closely related to the realities of the manufacturing environment. Such measures should focus on those areas that are critical to the success of operating in an AMT environment. Employees must also be made aware of what upper management feels are the critical areas so that they can focus their efforts on those variables. This communication can be accomplished through the budgeting and variance analysis procedures that are common in nearly all companies.

Performance measures also need to adopt a long-term perspective, which is the objective of most of the measures presented in this paper. At the same time, management compensation must be tied to the performance measures so that management has an incentive to achieve the goals that have been established at the time of the budget.

In reviewing the literature on product costing, some key points were common to many of the authors. With the increased competitiveness and shorter life cycles of products created by an AMT environment, the need for accurate product costing has never been greater. Through the use of techniques such as the two-stage procedure of cost allocation and activity-based costing, companies can feel more confident that the reported product costs are closer to truth than the costs developed under more traditional techniques.

New terms such as life-cycle costing, cost drivers and long-term variable costs have been introduced to the field of management accounting, and their meaning and uses were discussed. The impact of total costs moving away from direct- labor-based to overhead-based in an AMT environment was also highlighted. Using only volume-related measures, such as labor hours or machine hours, is not sufficient to properly account for the variety of activities that cause overhead to be incurred. AMT allows companies to offer a great deal of diversity in their product lines, and it is crucial that companies monitor and control the cost of complexity that results in such an environment. A growing number of companies have instituted some of the suggestions and improvements noted throughout this paper, but these companies represent the minority. Appendix A provides a listing of Harvard Business School cases that highlight many real world applications of the topics discussed in this paper.

5. RECOMMENDATIONS

In order to take full advantage of AMT, companies must first learn to simplify both their manufacturing process and their management accounting systems. Schonberger (1987) offers excellent advice on the notion of simplification of the manufacturing process. Once the manufacturing process is simplified, a company can begin, if desired and appropriate, to move toward the ultimate objective of AMT--a completely computer-integrated manufacturing facility.

The idea of simplifying an accounting process can also be thought of as appropriate in such an environment. This notion refers to more clearly focusing on what the objective of the management accounting function is, and not trying to have one accounting system serve a variety of needs. Once the different purposes of the MAS are defined, then a company can begin to use the MAS as a strategic tool and to supplement the benefits of adopting AMT.

A great deal of research still needs to be done in the area of accounting systems in an AMT environment. One area that has been virtually unexplored is that of the behavioral consequences of working and managing in the factory of the future. How will people respond to spending a significant part of their day interacting with machines instead of people? How will people respond to the pressure of being responsible for achieving zero defects? What about the impact of having one person responsible for the production of a given product, when that same product used to be handled by 10 people? Certainly these "people" issues will become more important as companies begin to implement AMT. For an interesting reference on this topic see Zuboff (1988).

Another fruitful area for research would be to explore the

bridge between performance evaluation and product costing. As an example, are some of the performance measures suggested earlier in the paper used as cost drivers as part of product costing? Also, are the performance measures incorporated into variance analysis or executive compensation plans?

The issue of education is also an area that needs to be explored. What is the proper type of training for an individual who would like to be an accountant for a company that is on the leading edge of technology? Should it be the traditional accounting program, with its emphasis on financial accounting? Or should it be more of a multi-disciplinary approach, with students taking courses in accounting as well fields such as production, engineering, computer technology and information systems? While the latter approach may seem to be ideal, questions of practicality arise. Who would choose such a major, who would teach in such a program and how many schools would be able to offer the appropriate courses?

More field research also needs to be done in order to discover what problems firms are encountering as they move into an AMT environment and how the company's MAS is responding to those problems. Along with this need, there is a demand for knowledge about the success stories. How have firms developed effective ways to account for the benefits of operating with advanced technologies, and how can these methods be applied to other firms?

As evidenced by this literature review, a great deal has already been written about the new performance measures that are needed in an AMT workplace. What is needed now is a critical appraisal of implementation issues. What are the potential pitfalls of using these new performance measures and what are the organizational politics involved in changing from traditional performance measures to the ones required by AMT? In order to examine these issues closer, it is necessary to go out and gather "war stories" from the companies that have met with success in these areas, as well as with companies that have run into problems along the way.

It is both an exciting and critical time for management accountants. It is exciting to be on the leading edge of technology and to be part of the team that helps a firm successfully implement these technologies. Yet it is also critical that accountants take advantage of their relative strengths, such as in measuring performance and product costing, before managers look elsewhere for the information necessary to enable their companies to compete successfully in the AMT environment.

6. REFERENCES

Bennett, Robert E., James A. Hendricks, David E. Keys and Edward J. Rudnicki. (1987). **Cost Accounting for Factory Automation.** Montvale, N.J.: National Association of Accountants.

Brimson, James A. (1986). "How Advanced Manufacturing Technologies Are Reshaping Cost Management," **Management Accounting**, March, 25-29.

Church, A.H. (1917). **Manufacturing Costs and Accounts.** New York: McGraw Hill.

Cooper, Robin. (1987a). "The Two Stage Procedure in Cost Accounting-Part One," **Journal of Cost Management**, Spring, 45-49.

Cooper, Robin. (1987b). "The Two Stage Procedure in Cost Accounting-Part Two," **Journal of Cost Management**, Summer, 43-51.

Cooper, Robin. (1988a). "The Rise of Activity Based Costing-Part One: What is an Activity-Based Cost System?" **Journal of Cost Management**, Summer, 45-54.

Cooper, Robin. (1988b). "The Rise of Activity Based Costing-Part Two: When Do I Need an Activity-Based Cost System?" **Journal of Cost Management**, Fall, 41-48.

Cooper, Robin. (1989a). "The Rise of Activity Based Costing-Part Three: How Many Cost Drivers Do You Need?" **Journal of Cost Management**, Winter, 34-46.

Cooper, Robin. (1989b). "The Rise of Activity Based Costing-Part Four: What Do Activity-Based Cost Systems Look Like?" **Journal of Cost Management**, Spring, 38-49.

Cooper, Robin and Robert S. Kaplan. (1987). "How Cost Accounting Systematically Distorts Product Costs," in **Accounting and Management: Field Study Perspectives**, eds. William S. Bruns Jr. and Robert S. Kaplan. Boston: Harvard Business School Press.

Goldhar, Joel D. and Mariann Jelinek. (1983). "Plan for Economies of Scope," **Harvard Business Review**, November-December, 141-148.

Hayes, Robert H. (1981). "Why Japanese Factories Work," **Harvard Business Review**, July-August, 57-66.

Hayes, Robert H. and William J. Abernathy. (1980). "Managing Our Way to Economic Decline," **Harvard Business Review**, July-August, 67-77.

Howell, Robert A., James D. Brown, Stephen R. Soucy and Allen H. Seed. (1987). **Management Accounting in the New Manufacturing Environment: Current Cost Management Practice in Automated (Advanced) Manufacturing Environments.** Montvale, N.J.: National Association of Accountants.

Howell, Robert A. and Stephen R. Soucy. (1987a). "The New Manufacturing Environment: Major Trends for Management Accounting," **Management Accounting**, July, 21-27.

Howell, Robert A. and Stephen R. Soucy. (1987b). "Operating Controls in the New Manufacturing Environment," **Management Accounting**, October, 25-31.

Howell, Robert A. and Stephen R. Soucy. (1987c). "Cost Accounting in the New Manufacturing Environment," **Management Accounting**, August, 42-49.

Johannson, Hank. (1985). "The Revolution in Cost Accounting," **P&IM Review and APICS News**, January, 42-46.

Johnson, H. Thomas. (1987). "The Decline of Cost Management: A Reinterpretation of 20th Century Cost Accounting," **Journal of Cost Management**, Spring, 5-12.

Johnson, H. Thomas. (1988). "Activity-Based Information: A Blueprint for World-Class Management Accounting," **Management Accounting**, June, 23-30.

Johnson H. Thomas and Robert S. Kaplan. (1987). **Relevance Lost: The Rise and Fall of Management Accounting.** Boston: Harvard Business School Press.

Kaplan, Robert S. (1983). "Measuring Manufacturing Performance: A New Challenge for Managerial Accounting Research," **The Accounting Review**, October, 686-705.

Kaplan, Robert S. (1984). "Yesterday's Accounting Undermines Production," **Harvard Business Review**, July-August, 95-101.

Kaplan, Robert S. (1988). "One Cost System Isn't Enough," **Harvard Business Review**, January-February, 61-66.

McNair, C.J., William Mosconi and Thomas Norris. (1988). **Meeting the Technology Challenge: Cost Accounting in a JIT Environment.** Montvale, N.J.: National Association of Accountants.

Miller and Vollman. (1985). "The Hidden Factory," **Harvard Business Review**, September-October, 142-150.

Schonberger, Richard. (1982). **Japanese Manufacturing Techniques.** New York: The Free Press.

Schonberger, Richard J. (1987). "Frugal Manufacturing," **Harvard Business Review**, September-October, 95-100.

Seed, Allen H., III. (1984). "Cost Accounting in the Age of Robotics," **Management Accounting**, October, 39-43.

Skinner, Wickham. (1969). "Manufacturing-Missing Link in Corporate Strategy," **Harvard Business Review**, May-June, 136-145.

Skinner, Wickham. (1974). "The Focused Factory," **Harvard Business Review**, May-June, 113-121.

"Strategic Manufacturing," a series by **Industry Week**. (1988). Cleveland: Penton Publishing.

Takeuchi, Hirotaka. (1981). "Productivity: Learning from the Japanese," **California Management Review**, Summer, 5-19.

Wheelwright, Steven C. (1981). "Japan-Where Operations Really Are Strategic," **Harvard Business Review**, July-August, 67-74.

Turney, Peter B.B. (1989). "Using Activity Based Costing to Achieve Manufacturing Excellence," **Journal of Cost Management**, Summer, 23-31.

Zuboff, Shoshanna. (1988). **In the Age of the Smart Machine.** New York: Basic Books.

7. APPENDIX A. HARVARD BUSINESS SCHOOL CASES EMPHASIZING PRODUCT COSTING/PERFORMANCE MANAGEMENT

Camelback Communications (185-179): Illustrates the distortions that can arise when decisions to drop product lines are based on data from a traditional accounting system.

Fisher Technology (186-302): Through the use of LOTUS templates, provides a basic introduction to the problem of using labor hours to allocate overhead in an automated environment.

Ingersoll Milling Machine (186-189): Looks at cost accounting in conjunction with advanced manufacturing technology and discusses life-cycle product costing.

John Deere Component Works (187-107/108): Deals with activity-based costing; a good example of looking at different systems for different purposes.

Mayers Tap Inc. (185-111): Focus of this case is on designing a cost system. Through the use of sophisticated LOTUS 1-2-3 templates supplied with the case, allows a complete and integrated look at cost system design.

Mueller Lehmkuhl GmbH (187-048): Indicates the problems that occur when the market defines products in one manner and the accounting system treats the products in a different manner.

Schrader Bellows (186-272): Highlights the problems that the current traditional cost system is causing, looks at the development of a new activity-based cost system and the behavioral problems that arise in such systems. Also includes LOTUS templates for use in discussing the case.
Siemens Electric Motor Works (189-089/090): Another case that illustrates the value of activity-based costing.

Tektronix (188-142/143/144): Explores the use of cycle time as the cost driver to be used in allocating costs to products.

Union Pacific (186-176/177/178): Explores the basic history of cost and management accounting. The use of a service company, as well as the change of competitive environment from a regulated industry to a non-regulated one, offers a different perspective on the use of activity-based costing. Also points out the difference between operational control and product costing.

Winchell Lighting (187-073/074/075): Examines the effect of different distribution channels on product profitability. Illustrates the impact of non-manufacturing costs on product profitability.

COST CONTROL AND PERFORMANCE MEASUREMENT: A PROBLEM DIAGNOSIS AND SOME RECOMMENDATIONS FOR THE NEW MANUFACTURING ENVIRONMENT

Lourdes D. Ferreira and Thomas W. Lin
School of Accounting
University of Southern California
Los Angeles, CA 90089-1421

ABSTRACT

The recent modernization and automation of American factories have created a demand for new cost management systems to make U.S. companies competitive in the world market. This paper summarizes the implications of the changing manufacturing environment on management accounting. It identifies major cost accounting problems in the areas of product costing, cost control and performance measurement. We review the major characteristics of traditional cost systems, such as emphasis on direct labor and unit volume, and contrast those with the current need to control activities that cause overhead costs to be incurred, such as inventory levels, production set-ups, reworks, downtime and other "cost drivers." Examples are presented on how the widespread practice of allocating overhead costs to products based on direct labor and other volume-related measures can distort product line comparisons and lead to wrong decisions with respect to strategically critical areas such as product costing, performance measurement and sourcing.

This paper suggests different criteria for allocating overhead to end products and proposes innovations for performance reports that compare actual vs. budget results. One of the recommended innovations relates to the inclusion of non-financial, operational variables in performance reports. Measures related to product and process quality and productivity are discussed. Examples of companies that have successfully implemented innovative product costing and performance measurement systems are also presented. Issues related to the design and implementation of performance measurement systems in modern manufacturing environments are also addressed. Some of the tradeoffs between simplicity and accuracy of the cost systems are discussed, and recommendations are offered about how the information necessary to update the cost systems can be obtained and included in regular performance reports.

1. INTRODUCTION

In the past decade, many American companies such as IBM, Hewlett-Packard, Allen-Bradley, Westinghouse, General Motors, General Electric, Eastman Kodak, Lockheed and Plessey, among others, have adopted advanced manufacturing technologies (AMT) in order to meet their global competition. The modernization and automation of U.S. factories should deliver increased quality and lower costs. However, unless companies modernize their cost management practices as well as their manufacturing processes, the benefits of these changes may never be realized.

This paper identifies major cost accounting problems in the areas of product costing, cost control and performance measurement. Suggestions are made as to how these problems can be resolved, and some implementation issues are discussed.

The recent professional and academic literature on new trends in management accounting suggests many reasons why companies need to revise their cost systems, but not enough specific recommendations have been developed. This paper presents recommendations and examples of new approaches in cost systems adopted by some innovative companies. These recommendations and examples are offered to provide guidance to those interested in revising their cost systems to meet the demands of new manufacturing environments.

This paper is organized in seven sections. The next section summarizes the impact of advanced manufacturing technologies on the changing manufacturing environment. Section 3 discusses major weaknesses of traditional cost accounting systems in the area of product costing. Section 4 examines how the internal reporting system for cost control can be modified to incorporate the changes identified in sections 2 and 3. Section 5 provides suggestions for financial and operational performance measures in the new manufacturing environment. Section 6 discusses some design and implementation issues, and a summary of the paper appears in the last section.

2. THE CHANGING MANUFACTURING ENVIRONMENT

Over the past decade, an increasing number of companies in the U.S. have been making changes in their manufacturing plants. Motivated by competitive pressures to reduce cost, increase productivity, improve quality and increase flexibility in response to customer needs, these companies have adopted innovations such as Just-In-Time (JIT) manufacturing, numerical control machines and robotics, Computer-Aided Design (CAD), Computer-Aided Manufacturing (CAM), Flexible Manufacturing Systems (FMS) and Computer-Integrated Manufacturing (CIM).

The major characteristics of modern manufacturing companies adopting the above innovations are high quality products and services, low inventories, high degrees of automation, faster throughput, greater flexibility and advanced information technology. These innovations have allowed firms to enjoy economies of scope and to diversify their product lines to meet unique consumer needs. But the new manufacturing technologies have also created a demand for increased coordination among various organizational units and for significant investments in production scheduling. These innovations shift the focus away from large production volumes necessary to absorb fixed overhead to a new emphasis on marketing efforts, engineering and product design. As a result, a world-class manufacturing company needs a world-class cost management system to produce high quality accounting information for more effective management decisions.

The list of adopted innovations in cost management and performance measurement systems is not so impressive. This situation is in part due to the fact that, in many companies, cost accounting is too often regarded as financial recordkeeping, serving the purposes of inventory valuation and product cost reporting, based on financial accounting standards. Managers in those companies may have ignored the potential benefits from updating their accounting systems. Even though obsolete cost accounting systems did not stop innovative companies from adopting competitive manufacturing systems, unrealistic cost information may have diminished the returns from those major investments in factory automation.

Kaplan (1984), in one of the first academic attempts to evaluate what changes were necessary in cost management systems, identified three major problems with traditional cost systems: (1) they distort product costs; (2) they do not produce the key non-financial data required for effective and efficient operations; and (3) the data they do produce reflect external reporting requirements far more than they do the reality of the new manufacturing environment. These problems are further discussed in the following sections, in which we mention major areas of weaknesses in applying a traditional cost accounting system to the factory automation environment.

3. PRODUCT COSTING

Traditional cost accounting systems currently do a very poor job in product costing because they no longer reflect how specific activities in the automated plant cause variations in major cost categories. Product costs have typically been monitored under three components: direct materials, direct labor and manufacturing overhead. Since traditional cost systems were developed when the labor component dominated total manufacturing costs, products requiring the highest labor input drove most of the production costs. Hence, the focus of these systems was on measuring and controlling direct labor costs.

Manufacturing overhead, defined as the sum of all production costs that cannot be identified directly with any product line (the sum of all indirect costs), was traditionally not a major cost element. It included expenses such as factory maintenance, utilities, insurance, supervisory salaries and so on. In the new manufacturing environment, overhead costs can become the major component of production costs because most of the expenditures associated with factory automation--new equipment depreciation and insurance, salaries for technicians, product engineers, research and development--are now included in the overhead account.

The percentage of total manufacturing costs related to direct labor has consistently decreased, with a corresponding increase in fixed overhead costs. For example, Horngren and Foster (1987) compare factory-overhead and direct labor percentages of total cost of goods sold for nine segments of the U.S. electronics industry, as indicated in Table 1 below. The segments with higher degrees of automation, such as semiconductors, exhibit a significantly higher proportion of manufacturing overhead costs relative to the more labor-intensive segments, such as software development.

Table 1. Percentages of Overhead and Direct Labor Costs

	Manufacturing Overhead	Direct Labor
Semiconductor	53.8%	20.9%
Active components	34.0	21.9
Electronic systems	33.3	23.9
Instruments	32.4	13.1
Production equipment	30.0	14.5
Passive components	29.8	21.6
Computer peripheral	27.5	6.5
Software	11.0	46.2

Source: Horngren and Foster (1987) p.445

In order to evaluate the profitability of different product lines, it has become even more necessary to do a proper allocation of overhead costs to end products, given the greater relative importance of overhead items. However, because overhead costs, by definition, are only indirectly related to end products, cost accountants have to devise some reasonable basis of "applying" overhead costs to individual products.

In traditional cost systems, overhead is often allocated to products on a basis of direct labor dollars or hours. Companies usually adopt a plant-wide overhead rate by dividing expected total overhead costs by total budgeted direct labor costs. Only a decade ago, overhead rates of 150 percent were fairly typical. Now overhead rates of 600 percent or even 1,000 percent are found in highly automated plants. As products move through a plant, they are charged overhead costs based on the plant-wide rate times the cost of direct labor required by each product. In traditional product costing systems, the number of units produced (unit volume) constitutes a major determinant of how overhead costs are incurred.

Such allocations in the factory automation environment can cause serious distortions. To allocate overhead based on direct labor cost, one has to assume that products with higher direct labor contents are responsible for greater overhead costs, leading them to be "charged" higher overhead allocations. Since this assumption is not consistent with automated environments, overhead allocations based on direct labor systematically overstate the costs of products with high direct labor content and understate the costs of other products that utilize more automated processes. In addition, because fixed manufacturing costs are now shared by several groups of products, allocation of manufacturing overhead to any individual product becomes more difficult.

Traditional techniques for product costing can cause distorted inventory valuation (even though the valuation is perfectly within financial accounting standards), wrong product line decisions, unrealistic pricing and ineffective resource allocations. Kaplan (1986) reports several dysfunctional consequences of using direct-labor hours as an overhead allocation base in machine-paced manufacturing environments.

Under new cost management systems, factory overhead costs should be allocated to products using cause-and-effect criteria. When production is highly mechanized, factory overhead costs, such as depreciation, supplies usage and indirect labor costs, are more closely related to machine utilization than direct labor usage. Therefore, machine-hours could be adopted as the base for allocating these categories of overhead costs. Cost systems based on cause-and-effect allocations tend to use multiple bases to allocate overhead costs, depending on the activity identified as the **cost driver** for each category of overhead costs. Such innovative costing systems are often referred to in the recent accounting literature as **activity-based costing** or **transaction costing** systems (Cooper and Kaplan, 1987).

In Figure 1, the two stages necessary for allocation of overhead costs are illustrated. The first stage refers to the

Figure 1. Two-Stage Overhead Allocation in an Activity-Based Cost System

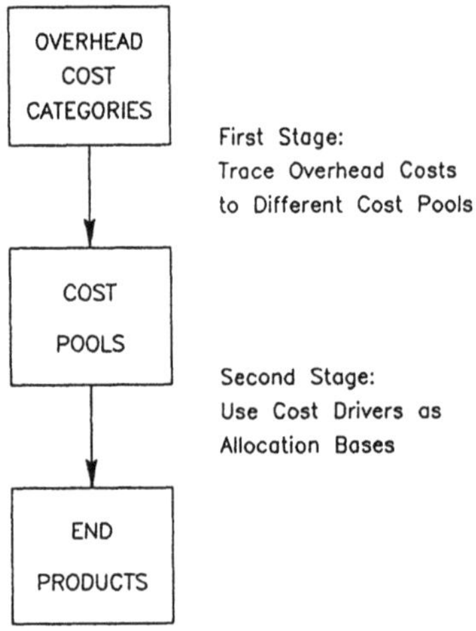

identification of the different activities conducted in cost centers (productive and service departments) that cause overhead costs to be incurred in these departments. This stage results in the allocation of each overhead cost category to a **cost pool**, which is defined as a set of activities that have the same cost driver or causal relationship with a certain category of overhead cost. For example, the costs associated with set-up can be identified as a cost pool.

The second allocation stage refers to the application of overhead costs from cost pools to end products, based on a measure of the cost driver or activity responsible for the incurrence of overhead costs in that cost pool. Continuing the previous example, as end products use up resources from the set-up cost pool, an allocation basis must be chosen for applying costs of that pool down to the product level. One possible allocation basis is the set-up time. Products that require longer set-up times are thus charged a higher proportion of the costs of the set-up cost pool than products that require shorter set-up times.

A critical difference between activity-based costing systems and traditional ones is that the allocation basis selected is the factor or activity that best explains how a certain category of overhead cost is incurred. Another unique characteristic of activity-based costing systems is that the activity basis selected for the second allocation stage, in which costs from the cost pools are applied to the end products, does not necessarily relate to unit volume. Set-up costs, for example, can be allocated based on the number of production runs for a particular product, independent of the total number of individual units produced. Table 2 summarizes the main differences between traditional product costing systems and innovative or activity-based costing.

Table 2. Traditional and Innovative Product Costing Systems

	Traditional Systems	**Innovative Systems**
Allocation basis	Direct labor dollars	Multiple cost drivers
Underlying assumption	Overhead costs vary directly with labor costs or unit volume	Each overhead cost category varies directly with its cost driver (independent of unit volume)

Companies are starting to adopt innovative allocation bases. For example, Hakala (1985) reports about a firm that uses a machine-hours allocation base. Here detailed records about machine usage, complete with distinctions between idle time, set-up time and operating time are kept for each machine. Schwarzbach and Vangermeersch (1983) report that one company uses a fourth cost pool category (in addition to direct-material, direct-labor and factory-overhead costs) termed "machining costs." This company keeps separate records for each of the key machines in its plant that include details about machine-operating time in minutes, energy usage and labor-hours of machine operators associated with their use.

Companies are starting to adopt innovative allocation bases. Hakala (1985) reports about a firm that uses a machine-hours allocation base, keeping detailed records on machine usage, complete with distinctions between idle time, set-up time, and operating time for each machine. Schwarzbach and Vangermeersch (1983) report that one company uses a fourth cost pool category (in addition to direct-material, direct-labor and factory-overhead costs) termed "machining costs." The company keeps separate records for each of the key machines in its plant that include details about machine-operating time in minutes, energy usage and labor-hours of machine operators associated with their use.

Some companies divide overhead costs into two categories--material overhead and conversion overhead. Material overhead includes the cost associated with planning, procuring, handling, storing, inspecting and distributing materials, as well as costs of engineering design. Conversion overhead includes supervision, payroll benefits and engineering support costs. While the number of parts can be used as the allocation base for material overhead, cycle time can be used to allocate conversion overhead.

The new emphasis on identifying the appropriate allocation basis for product costing is particularly necessary in the context of strategic decisions that will affect the long-run success of the firm. For example, if direct labor is used as the only basis for allocating overhead costs to products, product line decisions may be inadvertently made in favor of technology-intensive products and lead to the discontinuation of products with a high labor content. Management can be disappointed to see later that the traditional overhead allocations misrepresented the firm's new manufacturing technology and that the bulk of overhead costs (e.g., investments in automation) continued even after the labor-intensive product lines--which originally had "carried the burden" of allocated overhead--were discontinued. A simple numerical example in Table 3 illustrates how labor-based overhead allocations can

Table 3. Example of Distortion by Traditional Product Costing

	Product Line A (labor intensive)	Product Line B (technology intensive)
Direct materials	$ 100,000	$ 100,000
Direct labor	200,000	20,000
Actual overhead costs:		
General overhead	27,500	27,500
Technology overhead	20,000	200,000
Total manufacturing costs	347,500	347,500

Traditional cost system:

$$\frac{\text{Total overhead costs}}{\text{Total direct labor costs}} = \frac{275{,}000}{220{,}000} = \$1.25 \text{ per direct labor dollar}$$

Product costs based on direct labor allocation:

	Product Line A (labor intensive)	Product Line B (technology intensive)
Direct materials	$ 100,000	$ 100,000
Direct labor	200,000	20,000
Allocated overhead costs	250,000*	25,000**
Total manufacturing costs	550,000	145,000

* 1.25 times $200,000
** 1.25 times $20,000

misrepresent the use of overhead by product lines with different production technologies.

One of the most serious distortions caused by traditional product costing systems in the long run is that labor-based allocations can lead product managers to increase the use of external vendors for parts that require a labor-intensive process when produced internally. This outsourcing may even happen in cases when it would be actually more cost-effective to produce the part internally. Managers realize that a high number of direct labor hours would attract a high overhead charge, making labor-intensive parts or products seem to be cheaper if purchased from outside vendors, although in reality the part does not use that much overhead. This increased outsourcing leads to a loss in competitive domestic manufacturing, with further increases in the overhead costs of purchasing, receiving, handling and inspecting purchased parts (Horngren and Foster, 1987).

4. COST CONTROL

Traditional cost accounting systems control costs by preparing reports that compare budget-versus-actual amounts for each category of expense in a given department. As will be discussed in this section, these reports can lead managers to make the wrong decisions.

The emphasis on controlling volume and efficiency variances may encourage production department managers to keep labor and machines working to maximize output and efficiency, thus building inventory. The increase in inventory also allows managers to defer the portion of overhead costs that is applied to inventory, since these costs are reported as part of inventory in the balance sheet and are not included in the current income statement. Thus, in the short run, inventory building can help create favorable volume and efficiency variances, while at the same time overstating current earnings. Inventory buildup, however, is in sharp conflict with the desired productivity benefits associated with JIT and investments in AMT. Similarly, a purchasing department may order a lower quality material to achieve favorable purchase price variances. A production department may also pass on poor quality products to succeeding departments to avoid material usage variances.

If budget reports are prepared based on direct labor allocations, product managers will tend to pay excessive attention to controlling labor-hours instead of more costly categories such as materials and machining. Managers quickly realize that they can affect reported overhead costs by controlling direct-labor usage. However, this action does not control the actual incurrence of the larger materials and machining amounts. Johnson and Kaplan (1987, p.188) report that a plant spent 65 percent of computer costs processing information about direct labor transactions even though direct labor accounted for only 4 percent of total production costs.

Some authors now argue that traditional variance reports may have so many adverse consequences in the new manufacturing environment that it would be just better to eliminate them completely, especially since these reports require information that is costly to obtain, process and analyze. Given the limited benefits and arguable problems with variance reports based on control of labor costs, managers would be better off by simply scrapping their old cost control systems.

We consider these solutions too extreme. Variance reports can be made relevant by a careful revision of the bases used for allocating overhead costs or through the inclusion of operating variables such as quality controls and productivity measures. Standards can be generated for these variables through budget negotiations, and actual results can be compared against standards on a timely basis. In highly automated environments, the frequency of reporting may have to be daily, or even hourly, to allow for fast corrective actions. Flexible budgeting, whereby overhead budgets are adjusted to the level of production activity, can be implemented to allow better analysis and further control of overhead costs.

Even though managers cannot control costs of end products directly, they have control over the activities that incur overhead costs. As mentioned in the previous section, such activities are referred to as **cost drivers** in the recent accounting literature. Examples of cost drivers include engineering change orders, space utilization, process changes, downtime, inventory levels or the number of vendors, products and parts. Table 4 presents examples of overhead cost drivers. These cost drivers should be included in regular reports and effectively monitored where appropriate.

The set-up reduction program at Harley-Davidson is a good example of how companies can control costs by focusing on the cost drivers (Schwind, 1985). Harley-Davidson was concerned that it had long and expensive set-up times. To avoid numerous set-ups, the firm had been producing large batches. However, since this solution was causing large inventories that tied up substantial

Table 4. Examples of Overhead Cost Drivers

Input	Output
Number of vendors	Number of products
Number of parts	Number of options or accessories
Number of engineering change orders	Number of orders
Incoming material lead time	Inventory levels
	Defect and scrap levels
	Number of customers
	Number of distribution channels

Process
Number of set-ups
Number of production schedule changes
Amount of rework
Space utilization
Downtime
Process changes
Length of manufacturing cycle

capital, the firm decided to launch a multi-year project to reduce set-up times. The project was so successful that some machines achieved zero set-up time, and overall set-up time was reduced by 75 percent. By controlling this major cost driver, the company could reduce lot sizes, reduce inventories and improve product quality.

One of the key areas of cost control emphasized by JIT manufacturing is the control over product and process quality. Total quality control (TQC) becomes essential to avoid severe costs of disrupting production processes operating with minimum inventories. Kaplan and Atkinson (1989) argue that, contrary to traditionally held belief, the "optimal" level of quality is "zero defects." Companies implementing quality improvement programs have realized that the costs of ensuring increased quality are more than compensated for by savings from reduced scrap, rework, warranty and other costs of lower quality. As a result, total manufacturing costs tend to decrease as quality levels approximate zero defects.

Morse, Roth, and Poston (1987) point out that, to achieve total quality control, investments need to be made in several areas--prevention, appraisal, internal failure and external failure. They give examples of companies, such as the North American Philips Consumer Electronics Corp., that report quality costs under these separate categories in order to better control them.

Table 5 presents examples of these four types of quality costs based on Morse (1983) and Clark (1985). The measurement of quality costs by separate categories helps managers select which ones yield the most gains in reducing total manufacturing costs.

Table 5. Four Types of Quality Costs

Prevention	Internal Failure
Preventive maintenance	Scrap
Quality engineering	Rework
Supervision	Reinspection
Quality training	Retest
Supplier assurance	Breakdown maintenance
Appraisal	External Failure
Quality control	Warranty
Inspection	Claims and allowances
Testing	Replacements
Supervision	Failure investigation
	Delinquent orders

Usually, investments in the preventive quality category, including quality training and product/process engineering, yield the best results. In automated factories, the earlier quality is "built into" the product, the more costs are saved. Hence, special attention is needed at the product design stages. Products designed with fewer parts and with simpler assembly end up requiring much less investment in the categories of internal or external failure.

In summary, cost controls in new manufacturing environments have to be revised to better focus managerial attention on the key activities responsible for the largest overhead cost categories. Table 6 summarizes recommendations for innovative cost control

Table 6. Cost Control in New Manufacturing Environments

Traditional Systems	Innovative Systems
• emphasis on overhead volume and efficiency variances	• emphasis on controlling cost-driver activities
• control of direct labor costs	• control of other operating variables (e.g., quality, inventory levels)
• frequency of internal reporting determined by external reporting requirements	• frequency of internal reporting determined by operating needs

systems and contrasts them with traditional cost control systems.

5. PERFORMANCE MEASUREMENT

One of the most powerful instruments of management control is the performance measurement system. The choice of measures usedto evaluate performance conveys information about what actions and results are expected or desirable. Simply put, "We manage what we measure." Since managers have only a limited supply of time, energy and skills to fulfill their responsibilities, it is important to use the performance measurement system to direct managers' attention to the most critical aspects of the business.

Given the significant changes in the manufacturing environment mentioned in the previous sections, one would expect that companies undergoing those changes would also implement new performance measures. However, this has not been the case (Johnson and Kaplan, 1987). Companies continue to use traditional financial measures that emphasize control of direct costs, such as contribution margin measures, or short-term measures of investment returns, such as annual return on investment ratios. These measures are arguably less relevant in the new manufacturing environment, if one considers that the bulk of factory automation investments cannot be traced directly to individual products or be evaluated in the short-term. Yet, even companies that have reached high degrees of factory automation still prepare elaborate variance analyses based on standards developed when direct labor was the critical cost element. The information provided by some of these traditional performance measures is at best useless, and, in many cases, it has hindered management from reaping the full benefits of manufacturing automation and electronic information processing.

Several recent studies, published in both the academic and professional literature, have suggested ways in which traditional performance measures should be revised to incorporate the effects of manufacturing innovations. In Tables 7 and 8, we summarize some of these suggestions, which are examined below.

5.1 Financial Measures

Table 7 shows some financial measures of performance, which suggests that such traditional measures will continue to be widely used. The new approach with respect to financial measures is to analyze performance in different dimensions, as pointed out under the performance analysis item in Table 7. For example, in a JIT

Table 7. Financial Performance Measures

Financial Results	Performance Analysis
Cost control	Products
Revenue generation	Customers
Profit	Location
Asset management	Business unit
Return on investment	
Return on assets	Budget vs. actual results
Residual income	Company vs. industry averages
Cash flow management	Company vs. major competitors

environment, the high degree of interdependence among operating units makes it meaningless to analyze performance of each individual unit as if they were stand-alone operations. Instead, we suggest that analyses of profits, return on investment or cash flow be prepared, aggregating the information for each product, customer, location or overall business unit, which comprises all related businesses.

The significant changes described in the previous section indicate that, in the new manufacturing environment, managers have much less control over the final output or results from each period. In any given month, for instance, most production costs are fixed (they do not vary with production volume), sunk (no decision can be made currently that would change these costs) and indirect (reflecting shared resources among different product lines or operating units).

This lack of controllability has to be clearly considered when evaluating performance, or undesirable outcomes may occur. For example, if output from a certain operating unit B is less than expected, it may be because another unit A did not provide the necessary inputs, or because downstream unit C was not ready to receive that much output from B. In such a situation, manager B may have done the right thing from the point of view of the organization as a whole because the manager decided not to source inputs from outside nor create unnecessary work-in-process inventory just to meet a production budget. In order to avoid creating incentives for managers to do what is not in the best interest of the organization, the performance measures need to take into consideration these interdependencies.

Among the financial measures in Table 7, probably the most difficult to implement in the new manufacturing environment are the asset management measures because of problems in allocating assets and investment to each segment of the organization. It is reasonable to argue that the costs associated with asset utilization, such as the costs related to new equipment purchase, financing costs and so on, are enormous and that they need to be considered in order to assess "true" economic performance. However, these costs can rarely be traced to particular product lines or operating units, since they are usually decided at corporate level and benefit several segments of the organization simultaneously.

In companies adopting flexible manufacturing systems or computer-integrated manufacturing, the amount of shared resources is rather significant. It thus becomes more difficult or unrealistic to allocate asset utilization to individual segments, and the information provided by return on asset or return on investment figures loses much of its relevance. In particular, if the objective of the performance evaluation exercise is to assess performance of a particular manager for purposes such as bonus payments or promotions, then it becomes even less important to include asset utilization measures that reflect decisions made by other managers higher in the hierarchy.

It may be helpful, however, to allocate non-traceable costs, such as corporate expenses for evaluating unit managers, when top managers want to enlist cooperation of the unit managers to reduce these corporate expenses (Johnson and Kaplan, 1987). This argument reinforces the idea mentioned above that, "We manage what we measure." Once unit managers see a charge for new product development or for interest expenses accruing from new equipment purchases in their profit-and-loss statements, they will be motivated to help corporate keep these costs under certain limits.

Marketing expenses are an example of corporate overhead that could be allocated to product lines to aid in the evaluation of the profitability of different distribution channels. Cooper and Kaplan (1988) report that many companies now spend more than 20 percent of total revenues for selling, general and administrative expenses. Companies that are constantly introducing innovative products need to incorporate these substantial marketing costs into the performance evaluation system so that proper resource allocation decisions can be made.

The performance analysis criteria listed in Table 7 also include external measures such as industry averages and competitive averages as benchmarks for evaluating financial performance dimensions. This suggestion is emphasized by Keegan, Eiler and Jones (1989), who warn that internally-driven measures may not provide the signals when the firm is losing its competitive edge. In practice, this recommendation requires a lot of effort in the gathering of industry and competitive data. Some of this data is available through trade associations or journals, but care must be taken so that the appropriate "peer group" of companies is selected for assessing comparative performance. For example, a company should compare its profitability and return performance with those of companies with similar degrees of automation, vertical integration, size, product portfolio and so forth.

5.2 Non-financial Measures

Apart from financial measures, companies in the new manufacturing environment need to collect, report and analyze data about key operational factors. In Table 8, we list some examples of non-financial measures in the areas of quality and productivity management, based on Howell and Soucy (1987b, 1988).

Table 8. Operational Performance Measures

Productivity	Quality
Inventory turnover rate	Prevention
Output rate	Appraisal
Equipment utilization/capacity	Internal failure
Equipment availability/downtime	External failure
Cycle time	
Average set-up time	
Waste time	
Production backlog	
Output per employee	
Output per salary dollar	
Output per direct labor hour	
Throughput rate	
Material yield	
Distance traveled by products	
Factory space reduction	

By focusing on these operational measures, managers will be able to influence the major cost categories that are affected by these cost drivers. Cost accountants and others involved in the selection of these measures must intimately understand the production processes so as to capture the cause-and-effect relationships between the aspects being measured and the final costs incurred (Howell and Soucy, 1987a).

5.3 Recommendations for Selection of Performance Measures

The task of selecting and implementing performance measures is obviously not easy. But for a measurement system to fulfill its purpose of motivating responses from managers, it is critical that the measures be kept simple and understandable. One also needs to resolve conflicts among various measures by establishing clear priorities. For example, if a manager has to decide between meeting a certain target for quality or a target for throughput time, the performance measurement system should make clear which aspect is most important for the company, so that the appropriate tradeoff is made.

It may be necessary to limit the number of measures used each time, and to change the chosen measures periodically, as warranted. The flexibility provided by automated management information systems allows frequent adjustments in the measures at minimum information processing costs.

Perhaps the most important recommendation with respect to the choice of performance measures comes from their link to the company's strategy: the measures should be derived directly from the strategies chosen and be consistent with the long-term organizational goals. For instance, if a company adopts a strategy of focusing on products with short life cycles, the key success factor for this company becomes the constant introduction of innovative products capable of meeting consumer needs in specialized market niches. In this case, measures intended to control short-term marketing research costs or product development expenditures can be counter-productive. It may well happen that such a company is able to command premium prices for its special products based on their perceived value to the customers, and cost control is not an issue. IBM is an example of a company adopting such strategies, and it adopted performance measures to assess achievements in product development, as well as in other areas not directly related to cost control.

Keegan, Eiler and Jones (1989, p. 48) argue that, "Measurements will either cause strategy to be implemented or frustrated." They propose that performance measures have vertical and horizontal dimensions--the measures need to be consistent with the company's overall strategy from the top of the hierarchy down to all operating levels, and they also need to reflect the various functions in the organization to promote "horizontal" integration. Performance measures should be tailored to the different demands placed on different organizational units.

Such differentiation has limits, however, because of issues of horizontal equity. For instance, in a large diversified chemical processing company, although managers of mature product lines may be required to generate enormous volumes of cash flow, managers of emerging products may be allowed to maintain negative net cash flows while focusing on product development. If the managers who are under corporate pressure to generate large positive cash flows perceive such a system to be unfair, more homogeneous measures may be recommended.

In Table 9, we summarize some of the recommendations about how performance measures can be adapted to the realities of advanced manufacturing environments. Although these recommendations are general enough to be applicable to different

Table 9. Summary of Recommendations for Performance Measures

- recognize interdependence among operating units
- avoid including items outside control of the manager in performance reports
- collect, report and analyze measures of product/process quality and productivity
- keep the measurement system as simple as possible and establish priorities among various performance measures
- revise and refine the performance measures as warranted

organizational settings, it is practically impossible to provide comprehensive lists of performance measures without considering the specific technology, products, customers and processes involved in each situation.

Furthermore, performance measurement systems, even the most advanced and sophisticated ones, cannot substitute for managerial talent and judgment. Certain intangible, and yet critical, aspects of the business, such as the motivation and skill level of the employees, customer satisfaction and the quality of the supplier relationships, will still remain beyond measurement. These aspects are, in fact, directly related to the firm's competitive advantages (Porter, 1985). They deserve particular attention and regular assessments, especially in industry environments with fast-paced changes.

6. DESIGN AND IMPLEMENTATION ISSUES

Many of the cost management systems and performance measures currently used by automated manufacturing companies still suffer from some of the deficiencies described previously. Some systems simply automated the reporting activities originally done manually, with little or no changes, and with no consideration of the new manufacturing realities. For example, companies that manufacture products with short life cycles continue to use monthly reports comparing actual vs. budget amounts for several expenditure items. In these companies, most of the production effort is dedicated to continuous development of new products. Major upfront investments are necessary to develop innovative products, and the cash inflows from new product introduction usually come only months or years after the development expenditures. Yet, these companies ignore the fact that monthly reports do not represent an adequate match between costs and revenues in a given month.

When designing performance measurement systems for such companies, it would be more appropriate to keep a long-term focus and to control costs and revenues over the lifetime of the product, instead of preparing monthly reports. As illustrated in Figure 2, major initial cash outlays can be matched with revenues

Figure 2. An Example of Costs and Revenues from a New Product Introduction

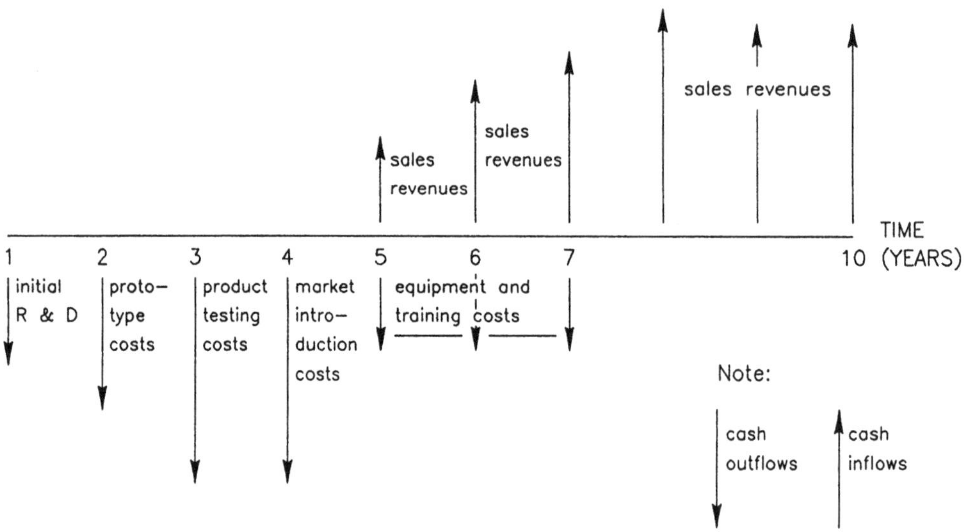

of multiple future periods, recognizing that costs such as research and development, new equipment acquisition, or investments in human capital (training, recruiting, etc.) benefit future cash flows. Rather than trying to artificially allocate these costs over several months, it would be better to recognize the long-term cash flow patterns, and accumulate the cash flows from the pre-production to the post-production stages. In manufacturing settings with short life cycle products, variable costs represent only a small percentage of total production costs, and the traceable portion of costs such as administrative and marketing expenses can be tracked over the product life cycle. From a managerial standpoint, information on the long-term net cash flows from each new product is more relevant than monthly reports of cash inflows and outflows arbitrarily allocated to that period.

Cost and management accountants need now to cooperate with engineers and manufacturing managers to design new cost management systems for factory automation. Cost drivers need to be identified, quality costs should be measured, and non-financial performance indicators are required. Yet the basic information necessary to implement these changes is not usually available, because most companies do not bother to collect it.

To obtain the necessary information for new cost management and measurement systems, the people directly involved in operating activities will probably need to be interviewed. Each operating and support department should be carefully studied to analyze its multiple activities so that cost pools in which the homogeneous cost drivers that are responsible for each cost category can be identified. For example, a quality control department may have three cost pools: inspection of incoming materials (cost driver: number of purchase orders), inspection of work-in-process (cost driver: number of set-ups), and inspection of finished goods (cost driver: cost of goods sold).

Understanding the production process and identifying cost drivers requires a significant amount of persistent effort. Depending on the particular circumstances, a major overhaul may be necessary to update the cost control and performance measurement systems. Such drastic changes should not be required frequently, though. After the main changes are introduced, companies can update only certain aspects of the cost and performance measurement systems periodically as changes in product mix, technology or customer base cause the current systems to become obsolete. The effort to redesign cost systems is mostly rewarded when there is high product diversity, various cost drivers, multiple channels of distribution and a wide range of batch sizes. In some cases, companies may decide that their particular manufacturing environment does not require further sophistication in the cost and performance measurement systems.

Before new cost systems are formally implemented in complex manufacturing environments, the new reports and measures should be maintained in parallel with the old systems for a trial period of perhaps one year. During this period, managers and accountants will have time to familiarize themselves with the new system, assess its comparative effectiveness, fine-tune it and train others to make use of the new information.

The availability of computerized systems may make the implementation of new cost control and performance measures much simpler than expected. Numerical control machines, for example, usually already store the operational data necessary to implement detailed monitoring of productivity measures. Electronic data processing also facilitates the adoption of multiple cost pools, cost drivers, and frequent changes in the assumptions in the cost management system. Furthermore, a manager may prepare his or her own customized report using a personal computer linked to the central computer in which the raw operational data is stored.

The sophistication and flexibility made possible by today's information processing capabilities can be overwhelming, though. Special attention is necessary to keep the performance measurement and cost control systems simple and objective. Yet, although the challenge of capturing the complexities of the current manufacturing environment into the accounting systems is great, the task is unavoidable. The price of not accepting this challenge may be as high as organizational survival.

7. SUMMARY

The modernization and automation of U.S. factories should deliver increased quality and lower costs if managers and accountants will cooperate to develop new cost management systems for product costing, cost control and performance measurement. This paper has identified major weaknesses of traditional cost accounting systems. We have presented suggestions for new factory-overhead allocation bases and some additional operating measures that focus on quality, cost drivers, service and other key success factors. We have also addressed design and implementation issues such as how to collect the necessary operational information, how to prepare performance reports with a long-term focus and how to introduce new cost and performance measurement systems.

8. REFERENCES

Clark, John. (1985). "Costing for Quality at Celanese," **Management Accounting**, March, 42-46.

Cooper, Robin and Robert S. Kaplan. (1987). "How Cost Accounting Systematically Distorts Product Costs," in **Accounting and**

Management: Field Study Perspectives, W. Bruns and R. Kaplan (ed.), Boston: Harvard Business School Press, 204-228.

Cooper, Robin and Robert S. Kaplan. (1988). "Measure Costs Right: Make the Right Decisions," **Harvard Business Review**, September-October, 96-103.

Hakala, Gregory. (1985). "Measuring Costs With Machine Hours," **Management Accounting**, October, 57-61.

Horngren, Charles. T. and George Foster. (1987). **Cost Accounting: A Managerial Emphasis.** Englewood Cliffs, NJ: Prentice-Hall (Sixth Edition).

Howell, Robert A. and Stephen R. Soucy. (1987a). "The New Manufacturing Environment: Major Trends for Management Accounting," **Management Accounting**, July, 21-27.

Howell, Robert A. and Stephen R. Soucy. (1987b). "Operating Controls in the New Manufacturing Environment," **Management Accounting**, October, 25-31.

Howell, Robert A. and Stephen R. Soucy. (1988). "Management Reporting in the New Manufacturing Environment," **Management Accounting**, February, 22-29.

Johnson, H. Thomas and Robert S. Kaplan. (1987). **Relevance Lost: The Rise and Fall of Management Accounting.** Boston: Harvard Business School Press.

Kaplan, Robert S. (1984). "Yesterday's Accounting Undermines Production," **Harvard Business Review**, July-August, 95-101.

Kaplan, Robert S. (1986). "Accounting Lag: The Obsolescence of Cost Accounting Systems," **California Management Review**, Winter, 174-199.

Kaplan, Robert S. and Anthony A. Atkinson. (1989). **Advanced Management Accounting.** Englewood Cliffs, NJ: Prentice Hall (Second Edition).

Keegan, Daniel, Robert Eiler and Charles Jones. (1989). "Are Your Performance Measures Obsolete?" **Management Accounting**, June, 45-50.

Morse, Wayne J. (1983). "Measuring Quality Costs," **Cost and Management**, July-August, 16-20.

Morse, Wayne, Harold Roth and Kay Poston. (1987). **Measuring, Planning, and Controlling Quality Costs.** Montvale, NJ: National Association of Accountants.

Porter, Michael E. (1985). **Competitive Advantage.** New York: Free Press.

Schwarzbach, Henry and Richard Vangermeersch. (1983). "Why We Should Account for the 4th Cost of Manufacturing," **Management**

Accounting, July, 24-28.

Schwind, Gene. (1985). "Harley-Davidson Explains the Techniques of Revitalization," **Material Handling Engineering**, February, 174-178.

MANAGEMENT ACCOUNTING AND THE ADOPTION OF FLEXIBLE AUTOMATION

Tony P. Dimnik
School of Business Administration
University of Western Ontario
London, Ontario, N6A 3K7, Canada

ABSTRACT

There is a widely held but little tested supposition that management accounting and control systems are retarding the adoption of flexible automation technologies such as robots and other computer controlled machines. This chapter describes a Canadian study that examines the relationship between adoption of flexible automation and those characteristics and uses of accounting systems most often cited as inhibiting adoption: emphasis on budget performance for managerial evaluation, time horizons of budget targets and reports, time horizons of capital budgeting criteria, emphasis on financial criteria in justifying capital investments, and difficulty in quantifying costs and benefits of automation. Data for the Canadian study were collected by a mail questionnaire survey and personal interviews of 32 managers of plants supplying parts to General Motors of Canada. The study found that the time horizon of budget targets and reports was the only accounting variable significantly correlated with adoption. The chapter concludes with the recommendation that North American management wishing to promote the adoption of advanced technologies lengthen the time horizons of budget targets and reports.

1. INTRODUCTION

There is a widely held opinion that management accounting and control systems (MACS) are retarding the adoption of flexible automation (FA). Industry leaders and authors have made increasingly forceful statements that traditional accounting-based methodologies are inadequate for evaluating and justifying investments in FA (Canada, 1986). Existing accounting-based performance measurement systems are also said to be inadequate and hindering the transition to the organization and technology required for the "new industrial competition" (Kaplan, 1984). Accountants are being urged to rethink their costing and performance evaluation methods because "yesterday's systems cannot be allowed to be a brake on today's--let alone tomorrow's--technological advances" (Sheridan, 1986). The challenge, we are being told, is to devise new, non-financial, longer-term, measures of managerial performance and new, non-financial measures of the benefits of advanced technologies (Kaplan, 1983). Surprisingly, such ubiquitous and strongly held views have been rarely tested by empirical research.

This chapter is organized as follows. Section 2 discusses current criticisms of MACS and studies of MACS and FA adoption. Section 3 describes a study of MACS and FA adoption in 32 Canadian manufacturing plants. Section 4 examines evidence from the Canadian study that suggests FA adoption is related to the time horizons of budget targets and reports. Finally, Section 5 proposes some future research projects and advises senior executives who wish to promote the adoption of FA in their firms to increase the time horizons of budget targets and reports.

2. THE LITERATURE

The role of MACS in the adoption of FA has been discussed in the practitioner-oriented and academic literatures of three disciplines: general management/strategic planning (e.g., **Harvard Business Review, Journal of Business Strategy**), production/engineering (e.g., **Production, The Engineering Economist**), and accounting (e.g., **Management Accounting, The Accounting Review**). A review of this literature can be found in Finnie (1988).

Without exception, the literature is critical of the role of MACS in the adoption of FA. Writers in the management stream complain of the use of short-term financial targets to evaluate managers and the use of analytic techniques like discounted cash flow for strategic investments (Hayes and Abernathy, 1980). To promote adoption of FA they prescribe lengthening the time horizons of budget targets and reports (Sheridan, 1986), lowering hurdle rates (Hayes and Garvin, 1982; Pearson, 1986), and using non-financial methods of evaluating proposed investments in advanced technologies (Bylinsky, 1986). Similar criticisms and prescriptions are found in the accounting literature (Kaplan, 1983, 1984, 1985; McLean 1986; Merchant and Bruns, 1986). The tone of the production/engineering literature is more insistent. Some production managers call for the abandonment of traditional financial justification methods (Huber, 1985). "Rather than relying on some type of 'creative' accounting procedure to reduce all the intangibles to dollars and cents", they advocate the use of non-accounting data and non-traditional techniques of analysis (Bernard, 1986). Engineering academics have responded to these criticisms by suggesting improvements and alternatives to traditional justification methods (Blank, 1985).

To sum up, the literature advises senior executives of manufacturing firms who wish to encourage investment in FA to place less emphasis on budget performance for evaluating managers, to increase the time horizons of budget targets and reports, to lower hurdle rates and/or lengthen payback thresholds, to place less emphasis on financial criteria in evaluating investments in FA, and to use non-financial criteria to evaluate the costs and benefits of FA. This chapter reports on a study that empirically tested the validity of these prescriptions.

Only a few previous studies have explored the relationships between accounting and investment in advanced manufacturing technologies. In one study, Woods et al. (1985) looked at the capital budgeting procedures of 92 U.K. mechanical engineering firms that had considered investing in CAD/CAM technologies. Although adopters generally made greater use of traditional financial justification methods such as payback and discounted cash flow, it is unclear whether adopters and non-adopters differed in their methods of analyzing FA proposals. In another study, Schwarzbach (1985) surveyed 112 North American manufacturing firms and found no demonstrable relationship between level of automation and the "sophistication of the accounting system." Finally, in the most comprehensive study of MACS and FA to date, Howell et al. (1987) mailed questionnaires to a sample of 1,000 preparers and 1,000 users of management accounting information, and to 217 other persons such as accounting academics. In addition to their mail survey, Howell et al. conducted on-site interviews with more than 100 financial/accounting and operating executives of 17 companies chosen because of their reputation for progressive manufacturing practices. The stated aim of their study was to provide a "baseline as to the present state of the art of management accounting," but Howell et al. seemed to be looking for evidence that there is something wrong with traditional accounting systems:

> Although dramatic changes are taking place in the marketplace and inside the factories, little seems to be happening in the areas of financial analysis, management accounting, and operating controls to support the changes. A feeling is growing throughout business that management accounting must adapt to the new manufacturing environment. (page vii)

Howell et al. found that the users of accounting information who responded to the mail questionnaire were especially dissatisfied with their MACS.

Neither the Howell et al. study, nor those that preceded it, directly addressed the validity of the prescriptions that promise to promote FA adoption. Indeed, no study has demonstrated the existence of <u>any</u> relationship between FA adoption and characteristics and uses of MACS. Such evidence is a prerequisite for prescriptive or normative purposes.

3. THE GMC STUDY

To investigate the role of MACS in FA adoption, a study was conducted of suppliers of parts to General Motors of Canada (GMC). The GMC study did not presume any relationship between MACS and FA adoption. It was an exploratory study that considered the basic question of whether or not there is any relationship between MACS and FA adoption.

The study was designed to address three sampling issues: firm heterogeneity, respondent heterogeneity, and self-selection bias.

These issues will be illustrated with references to the Howell et al. study. The first sampling issue addressed by the GMC study is firm heterogeneity. Howell et al. based their conclusions on data from firms in industries as diverse as automotive, aerospace, electronics and consumer products. Industry factors such as manufacturing process, product life cycle and competitive environment may confound relationships between MACS and FA adoption and should thus be controlled by statistical or sampling methods. The second sampling issue is variety in the hierarchical levels of respondents. A MACS must accommodate many different users, and while some users may find fault with a MACS, others may be satisfied with the system. The same characteristics and uses of a MACS found unsatisfactory by a plant manager, may be considered essential by a senior executive. Studies such as the one conducted by Howell et al. include a cross section of managers from all levels of their firms. A study of MACS and FA adoption should control for respondents' organizational positions. The third issue, self-selection of respondents, is always a concern in empirical research. Howell et al. received valid mail questionnaire responses from only 26% of preparers and 6.4% of users, and had no information on how respondents and their firms may have differed from non-respondents and their firms.

The GMC study, as have previous accounting studies, controlled for confounding variables by selecting a sample of firms from a single industry (Merchant, 1981). The Canadian automotive parts industry was chosen for several reasons. First, the industry includes a wide, but tractable, variety of manufacturing processes. The GMC sample included plants from four product groups: metal stamping, plastic molding and finishing, rubber parts manufacturing, and functional and decorative die casting. Thus, even though the GMC study was conducted within a single industry, the results can be generalized across several manufacturing processes. Second, the Canadian automotive parts industry is one of only a few that have sufficiently high levels of FA adoption to provide a meaningful test of the relationships between MACS and FA adoption (Economic Council of Canada, 1987). Third, the GMC sample represents a major segment of the North American economy. General Motors was the single most important customer of the surveyed plants, but, on average, the plants shipped 55% of their output to other assemblers. Because the Canadian and American automotive industries are integrated, the GMC sample is representative of the broader North American automotive parts industry.

The GMC study controlled for the hierarchical position of respondents by measuring FA adoption at the plant level and by surveying plant managers. FA adoption can be measured at either plant or firm (multi-plant) levels. Since FA adoption can vary widely among plants within the same firm, and since characteristics and uses of MACS can also differ from plant to plant in the same firm, it follows that FA adoption and MACS attributes should be measured at the plant level. Commensurate with the decision to measure plant level adoption was the decision to survey plant managers. Plant managers were defined as people who are responsible for the operation of a manufacturing plant and who make capital investment proposals to senior executives, but who do not have authority to approve major expenditures. Plant managers are often cited as being principal players in the adoption and utilization of FA (Bessant, 1982; Haka, 1987).

The GMC Director of Purchasing sent a personal invitation to the senior executives of 51 GMC parts manufacturers. The invitations asked the executives for permission to survey the managers of 51 specifically identified plants. Thirty-two, or 63%, of the firms agreed to participate in the study. To assess response bias and the possibility that "successful" firms were more likely to participate in the study than "less successful" ones, 26 GMC purchasing, quality assurance and engineering personnel were asked to rate all 51 suppliers on FA adoption and on overall performance. Comparisons of the GMC ratings of the 32 participants and 19 non-participants showed no evidence of self-selection bias on FA adoption (two-tailed t-test: $p=.74$) or performance (two-tailed t-test: $p=.51$).

The GMC study employed a two-wave survey method: mail questionnaires were sent to and returned by all 32 plant managers and, after the mail questionnaires had been returned, 29 of the plants were visited and their managers personally interviewed. Table 1 shows some of the plant and respondent characteristics of the GMC sample.

Table 1. Plant and Respondent Characteristics

The Respondents

	Mean	Standard Deviation	Minimum	Maximum
Age (in years)	40.8	7.4	30	61
Tenure with firm (in years)	9.5	8.6	1	26
Tenure as plant manager (in years)	3.6	2.8	1	10

The Plants

	Mean	Standard Deviation	Minimum	Maximum
Number of employees	326	309.8	30	1600

Annual Sales in Canadian Dollars	Percent of Plants
-under $5-million	6
-between $5 and $10-million	10
-between $10 and $20-million	28
-between $20 and $50-million	28
-over $50-million	28

Adoption of Flexible Automation Technologies	Percent of Plants
-robots	59
-programmable controllers	86
-CNC machines	10
-computerized material handling	24
-computer-aided testing	76
-CAD	55
-CAD/CAM capability	24

The remainder of this section describes the measurement of FA adoption and MACS attributes, and discusses the reliability and validity of the measures.

3.1 The Measures

FA was defined to include the following reprogrammable, computer-controlled manufacturing technologies: robots, computer numerical control machines, programmable controllers, automated materials handling equipment, computer-aided inspection and testing devices, and CAD/CAM. FA was measured by a self-rating mail questionnaire instrument that asked respondents to compare their plants FA adoption against their competitors and by an FA technology count conducted during the plant visits. The self-rating FA adoption variable was created by summing scores on four, five-point, Likert-scale items, such as: "Compared to your competitors, your plant's use of robots, CAD/CAM and other flexible automation manufacturing technologies is...(Circle number from 1=lowest to 5=highest)." The FA technology count was adjusted for plant size by dividing numbers of machines by number of plant workers. Thus, a plant with 5 robots, 10 programmable controllers, and 100 workers would score .15 on the FA technology count measure. The FA self-rating and count variables were highly correlated (Spearman correlation coefficient of .55, with a one-tailed significance level of p=.001).

The GMC study measured the five characteristics and uses of MACS most often cited as retarding the adoption of FA: emphasis on budget performance for managerial evaluation, time horizon of budget targets and reports, length of payback thresholds, emphasis on financial criteria in justifying investments in FA, and difficulty in quantifying costs and benefits of FA investments. With the exception of payback thresholds, which were measured with similar mail and personal interview instruments, each of the MACS attributes was measured in two different ways: with a multi-item mail questionnaire instrument and with a single item personal interview instrument.[1] Scores on the multi-item mail questionnaire instruments were summed to form variables. Face-to-face interviews were tape-recorded and responses were coded. Brief descriptions of each of the measures follow.

The mail questionnaire measure of the emphasis on budget performance for evaluating managers was a seven-item instrument from Merchant (1981) that assesses the extent to which budgeting systems are linked to external rewards. A sample item is: "Budget performance is an important factor in advancing my career...(Circle number from 1=strongly disagree to 5=strongly agree)." The personal interview version was Hopwood's (1972) evaluation style instrument which asks respondents to rank the importance of various performance measures in their evaluation by superiors. Responses were coded from 1 (no accounting measures of performance) to 4 (meeting budget targets is top ranked evaluation criterion).

The time horizon of budget targets and reports was measured by two mail questionnaire items asking respondents to indicate the longest time horizon of cost, sales, profit or ROI targets and the longest time period covered in statements of cost, sales, profits, ROI and/or variances from budget targets. Lorsch and Morse (1974) used a similar instrument to measure the "time dimension of formal practices." Scores on the budget targets item could range from 1 (daily targets) to 6 (targets for periods longer than one year) and scores on the budget reports item could range from 1 (daily reports) to 7 (reports for each project or contract). The corresponding interview version of the time horizon variable incorporated responses to questions about planning and budgeting horizons and was coded on a five-point scale where a score of 1 indicated a horizon of much less than one year and a score of 5 indicated a horizon of much more than one year.

The mail questionnaire measure of emphasis on financial criteria in the justification of investments consisted of four items of the type: "I can get new equipment only if I can convince my superiors that the investment makes financial sense...(Circle

---

[1] The mail questionnaire asked respondents to report hurdle rates but few of the plant managers used discounted cash flow techniques to analyze projects and so the study focused on payback thresholds.

number from 1=strongly disagree to 5=strongly agree)." The interview version focused on payback. Managers were asked to assess the importance of payback in judging investment proposals and responses were scored from 0 (payback not important) to 2 (payback very important).

The mail questionnaire measure of difficulty in quantifying the costs and benefits of FA consisted of two, five-point, Likert-scale, items asking respondents to agree or disagree to statements that it is difficult to estimate the costs and benefits of FA. Higher scores on this variable indicated greater difficulty in quantifying the costs and benefits of FA adoption. The corresponding personal interview question asked respondents if they felt their justification system frustrated efforts to obtain new technologies. Responses to this question, which were coded from 0 (no frustration) to 2 (frustrated in promoting projects), reflected both the difficulty in quantifying costs and benefits and the impact of constraints such as availability of capital.

3.2 Reliability and Validity

Several tests were conducted to assess the reliability and validity of the MACS measures. All multi-item measures exceeded minimum standards for internal reliability as assessed by Cronbach's alpha. To assess convergent validity, Spearman correlation coefficients were calculated for each pair of (mail and interview) measures. Coefficients and their one-tailed probabilities follow: budget emphasis (.30, p=.06), time horizon (.32, p=.05), payback criterion (.63, p=.001), emphasis on financial justification (.45, p=.01), difficulty in quantification (.15, p=.23). Considering that the mail questionnaire and personal interviews were conducted several weeks apart and that the data collection methods differed greatly, the correlations between alternate forms of budget emphasis, time horizon, payback criterion and emphasis on financial justification seem to exhibit sufficient convergent validity. However, the correlation between the two versions of "difficulty in quantification" is very low. The reason for this may be, as was noted above, that the mail questionnaire instrument measured difficulty in quantifying FA costs and benefits while the personal interview instrument measured respondents' general attitudes towards their justification systems. Table 2 shows summary statistics of the FA adoption and MACS variables.

4. RESULTS OF THE GMC STUDY

The substantive results of the GMC study will be presented in two parts. The first part discusses the correlations between the MACS and FA variables which are shown in Table 3. The second part discusses impressions and conclusions drawn from the plant visits and the interviews with the plant managers.

4.1 Correlations Between MACS and FA Variables

Table 3 shows that only four of the twenty correlations between MACS and FA adoption variables are significantly different from zero. Three of the significant correlations are associated with the time horizon of budget targets and reports. The interview version of the time horizon variable has a positive but not significant relationship with self-rated FA adoption, but the three other time horizon correlations suggest a strong relationship between time horizons of budget targets and reports and FA adoption. The only other significant correlation is between self-rated FA adoption and mail-questionnaire reported payback thresholds. However, the three other payback correlations provide no evidence of a relationship between payback thresholds and FA adoption.

In general, the signs of the correlations in Table 3 are consistent with the literature criticizing the role of MACS in FA adoption. As examples of consistent signs, consider the four correlations dealing with emphasis on budget performance for evaluation: all four correlations are negative, indicating that greater emphasis on budget performance is associated with lower levels of FA adoption. To sum up Table 3, emphasis on budget performance for evaluating mangers, length of payback thresholds,

Table 2. Summary Statistics of FA and MACS Variables

VARIABLES	MEAS. METHOD*	DESCRIPTION OF MEASURE	MEAN	STAND DEV.	ACT RANGE	POSS RANGE
Emphasis on budget performance for evaluation	MS	7 items -- 5 point scales	24	4.4	11-33	7-35
	PI	1 item -- 4 point scale	2.2	.98	1- 4	1- 4
Time horizon of budget targets and reports	MS	2 items -- 6 and 7 point scales	10	1.9	6-13	2-13
	PI	1 item -- 5 point scale	3.2	.99	1- 4	1- 5
Time horizon of payback thresholds	MS	1 item -- months payback	28	13.2	6-60	-
	PI	1 item -- months payback	27	11.8	12-60	-
Emphasis on financial criteria in justifying investments	MS	4 items -- 5 point scales	14	2.8	7-19	4-20
	PI	1 item -- 3 point scale	1.3	.59	0- 2	0- 2
Difficulty in quantifying costs and benefits of flexible automation	MS	2 items -- 5 point scale	5.6	1.8	2- 9	2-10
	PI	1 item -- 3 point scale	.48	.75	0- 2	0- 2
Adoption of flexible automation	MS	4 items -- 5 point scales	13	3.3	8-19	4-20
	PI	Machine count divided by number of workers	.08	.08	0-.33	-

* MS = Mail Survey; PI = Personal Interview

Table 3. Spearman Correlation Coefficients

CHARACTERISTICS AND USES OF MACS	MEASUREMENT METHOD FOR MACS VARIABLE	SELF-RATED FA ADOPTION (Mail Survey Measure)	COUNT OF FA ADOPTION (Interview Measure)
Emphasis on budget performance for evaluation	Mail	-.0508 N 31 p .786	-.2746 N 28 p .157
	Interview	-.2841 N 29 p .135	-.0705 N 29 p .716
Time horizon of budget targets and reports	Mail	.5601 * N 30 p .001	.5942 * N 27 p .001
	Interview	.2446 N 29 p .201	.5831 * N 29 p .001
Time horizon of payback thresholds	Mail	.4393 * N 27 p .022	.2537 N 24 p .232
	Interview	.1003 N 23 p .649	.0810 N 23 p .713
Emphasis on financial criteria in justifying investments	Mail	-.0700 N 31 p .708	.1190 N 28 p .546
	Interview	-.3148 N 27 p .110	-.0566 N 27 p .779
Difficulty in quantifying costs and benefits of flexible automation	Mail	-.3054 N 32 p .089	-.0326 N 29 p .867
	Interview	-.0972 N 27 p .630	.0425 N 27 p .833

* Significant at .05 level.

emphasis on financial criteria in investment justification and difficulty in quantifying costs and benefits of FA are not significantly correlated with FA adoption, but the time horizon of budget targets and reports is significantly correlated with FA adoption.

The relationships between MACS and FA adoption were explored in the interviews with the plant managers. Insights from these interviews are presented below and arranged according to each of the five characteristics and uses of MACS investigated by the study.

4.2.1 Emphasis on Budget Performance

Almost all respondents said they wanted more FA for their plants and most respondents claimed that the key to getting the FA they wanted was to have a good "track record." "Heavy hitters" or "managers with good batting averages" were more likely to apply for, and get, advanced technologies. By itself, emphasis on budget performance for evaluation may not be related to FA adoption. But the interaction between emphasis on budget performance and actual budget performance may be related to adoption. Consider a situation where budget performance is not used to evaluate managers. In such a situation it may be difficult for a manager to demonstrate competency. A history of good performance, as recorded by the accounting system, may make it easier for managers to get new machinery and equipment because they are able to demonstrate they have done well with what they have been given in the past. See Pinches (1982) for a discussion of how senior executives' perceptions of a project sponsor's record affect capital budgeting criteria.

The relationships among FA adoption, emphasis on budgets for evaluation, and managers' track records, may be affected by other contingencies. For example, the time span of performance measurement may be related to the development of track records. The automotive industry is cyclic in nature and the longer the time period over which evaluations are made, the better can senior executives assess the performance of plant managers. Of course, this assumes senior executives are sophisticated users of MACS. Based on the comments of plant managers, that would be a fair assumption.

4.2.2 Time Horizon of Budget Targets and Reports

The possibility that long-term budget targets and reports may assist managers in creating track records has just been noted. But any explanation of the strong correlation between budget time horizons and FA adoption must also consider managerial motivation and planning resources.

The plant managers in this study usually considered FA adoption in the context of implementing new programs (ie. bidding on and manufacturing new parts).[2] Managers with longer budget time horizons seemed to be more aware of, and concerned about, the economics of future production and were thus more likely to consider incorporating FA in new programs. For example, when considering a bid on a new part for GMC, one plant manager, whose firm had long-range budget targets and reports, discovered that the only way he could reconcile expected increases in labor costs with GMC's expectations of price decreases each year of a long-term contract, was to apply FA to the manufacturing of the part. The manager convinced senior executives of his firm to bid on the part using FA-based costs. His firm won the long-term contract, and adopted the FA. Competitors for the GMC contract had not addressed the issue of rising costs and falling prices. Their bids were based on simple extrapolations of current costs and current technologies.

Managers with longer budget time horizons had more resources for planning. Where long range budgeting was the norm, it was more acceptable for managers to commit time and money to the planning process. As one high-adopter put it: "You have to plan to have some time to plan." In this respondent's firm, lower-level managers held regular strategy and planning sessions on Saturday

---

[2] Many managers reported that FA adoption was successful only when new technologies were implemented in conjunction with new programs. Attempts to add FA to on-going programs often failed.

mornings. Process innovations were often discussed at these sessions. Contrast this culture with the one in many low-adopting firms where mangers spent little time making long-term plans and were not expected to "waste time" worrying about the distant future.

4.2.3 Payback

The fact that payback criteria were not related to FA adoption may be explained by the ubiquity of short-term payback criteria. Only 14% of respondents reported payback thresholds longer than three years. Managers cited three reasons for utilizing short payback periods in assessing potential investments in new technology. Preparing proposals, especially proposals for larger investments is time-consuming. As one manager put it, "I'm not going to spend hundreds of hours working on a proposal I know is going to be shot down."[3] To avoid wasting time, managers requested monies only for projects they knew would be approved: projects with quick paybacks. A second reason managers use very short payback criteria is their belief that senior executives are reluctant to continuously upgrade equipment:

> If you're not tough on your payback and your competition is, you'll be in trouble. If a better mousetrap comes along, you won't be able to get that better mousetrap because you've got this piece of equipment and you're waiting for your five year payback on it. So this new mousetrap's out there and you can't get it because you've got the old model and you still have to rationalize it.

A third reason for short paybacks is risk aversion. Several managers described personal strategies of avoiding projects with marginal paybacks. They stated they would not propose or support projects that marginally exceeded the formal payback criteria. In

[3] Managers' comments about supporting only those projects they know in advance will be approved confirm the view that MACS are especially important in the early (initiation) stages of capital budgeting (Bower, 1970; Pinches, 1982).

other words, given the choice, managers preferred to "go for the sure thing" and the "big winners."

It may be that short-term payback thresholds impede investment in advanced technologies, but for the reasons cited above, most managers in the GMC sample used very short payback criteria and the variable was not correlated with FA adoption. In samples with more variability in payback thresholds, payback may have a stronger relationship with FA adoption.

4.2.4 Emphasis on Financial Justification

Flexible automation technologies may be classified as stand-alone installations or integrated systems, and may be of an offensive or defensive nature (Bessant, 1982; Brealey et al. 1986). Stand-alone FA installations are single computer controlled machines that replace one function of a production line. Integrated FA systems are several computer controlled and coordinated machines that replace an entire line or automate an entire process. Most of the installations in this study were stand-alones and thus relatively simple and inexpensive. As a rough guide, the least expensive integrated system in the GMC sample cost about ten times as much as the average stand-alone installation.

In the personal interviews, mangers described another investment dichotomy: some investments are offensive and others are defensive. Here is how one manager distinguished between the two types:

> Defensive investments are those we have to make if we want to stay in the business we're in now. Offensive investments are for new business and we examine the payback on those very carefully.

An example of a defensive investment is the purchase of a robot to replace a human operator in a paint room because the room's environment does not meet government health and safety standards. An example of an offensive investment is the purchase of computerized equipment for testing a new part to convince customers that the firm is serious in its bid to manufacture the part.

In the GMC sample, financial justification seemed to be an issue only in those FA proposals that were of an offensive nature and that involved purchase of an integrated system. Investment proposals of a defensive nature, whether for integrated or stand-alone machinery, required little justification. Defensive investments had to be made if a firm was to stay in business. Offensive investments in stand-alone machinery were fairly straightforward and required little justification: such investments represented either a replacement of current machinery or a minor extension of current business. However, offensive investments in integrated FA were major strategic decisions and required much more effort to justify. It may be that emphasis on financial justification is a factor only in the consideration of offensive investments in integrated flexible automation. A worthwhile question for future study would be: is emphasis on financial justification related to the approval or rejection of integrated, offensive investments in FA? Research that attempts to answer this question will require a sample of plants whose management has considered offensive investments in integrated FA. The sample will have to include both managers that adopted the technology and managers that did not.

4.2.5 Difficulty in Quantifying Costs and Benefits

Managers in the GMC sample generally disagreed with the proposition that it is difficult to quantify the costs and benefits of flexible automation. The interviewees were quite happy with their MACS and with their accountants. That satisfaction can be partially explained by the fact that managers do not depend solely on their MACS to generate the numbers required to justify FA investments. For example, one manager ordered engineering time studies of a plastic molding operation and found large variances in the cycle time of the manually operated machines. The manager was able to justify an investment in robots by showing that the new technology would be more consistent and would lower the cycle time. He argued that shaving off a few seconds on each part could increase capacity to the point that the robots would pay for themselves in less than two years.

There is no way of knowing whether accounting numbers underestimate the benefits or overestimate the costs of FA projects that are judged uneconomical. However, if their comments are taken at face value, respondents had little difficulty quantifying the costs and benefits of FA. In general, respondents were able to quantify the benefits of the projects they wanted and the costs of those they opposed. What difficulties there were appeared to be less related to shortcomings in the MACS than to managerial commitment to a particular project. That is not to say that managers were able to justify every investment. For example, when management of a rubber parts plant could not justify an investment in an integrated FA system using their cost and benefit estimates, they hired outside consultants to evaluate the proposal. After tallying up revised estimates of costs and benefits, the consultants, who were also in the business of selling and installing FA, advised against the integrated FA system. Trying to duplicate the hand-eye co-ordination of human operators with computerized machinery was simply uneconomical for this rubber parts manufacturer.

A final comment on financial quantification concerns the role of accountants in the capital budgeting process. In many cases, respondents described their accountants as "proof readers." The accountants' role in capital budgeting was often limited to verifying that numbers in formal investment proposals were reasonable and consistent. Limited interaction between accountants and plant managers may explain why there was so little evidence of conflict between these two groups in regards to FA investments.

5. CONCLUSIONS

Several other research initiatives were suggested by observations from the face-to-face interviews.

Studies should be conducted to determine whether adopters of FA have "finer" MACS and are more sophisticated users of their MACS than low adopters. "Fineness" refers to the amount of detail in accounting reports. Although "fineness" of MACS was not measured in the GMC study, it appeared that the accounting reports of FA adopters categorized costs and revenues in smaller pools than did the reports of low adopters. It also appeared that FA adopters used the numbers generated by their MACS more than low adopters and that FA adopters were more sophisticated in their use of accounting numbers. Many of the FA adopters had formal training in accounting and well understood the potentials and pitfalls of their MACS. Several respondents commented that their knowledge of accounting had helped them acquire and manage their FA technologies.

Studies should also be conducted to determine how changes in strategy impact on MACS and FA adoption. Historically, many of the low adopters in the GMC sample had been managed as "cash cows." These plants had generated profits for decades but had received very little investment in new machinery and equipment. Wringing every last drop of profit from capital investments may have been an appropriate strategy for the automotive industry when it was in the mature phase of the product life cycle but the industry is now "dematuring" (Jones, 1985). Senior executives of several of the low adopters had recognized the need for more aggressive FA investment strategies. Research is required to see if the MACS of these firms, and others like them, change to accommodate new strategies and if changes in the MACS are necessary for the new strategies to succeed.

To sum up this chapter, there have been few studies that empirically tested the relationships between MACS and FA adoption. In contrast to studies which presumed that MACS are retarding investment in new technologies, the GMC study attempted to discover, which, if any, characteristics and uses of MACS are related to FA adoption. The GMC study surveyed 32 suppliers of parts to General Motors of Canada and found that the time horizon of budget targets and reports is the only MACS variable significantly correlated with FA adoption. The GMC results, taken as a whole, are somewhat surprising considering the widely held view that MACS are a serious impediment to FA adoption. However, the finding that budget time horizons impact on FA adoption supports the normative literature which recommends that senior executives employ accounting measures and procedures that reflect the long-term objectives of their firms. This advice implicitly assumes that budget time horizons "cause" FA adoption and are not "caused" by the adoption of new technology. While the GMC study does show that the time horizons of budget targets and reports are significantly correlated with FA adoption, inferences about causality cannot be drawn from simple correlations. However, interviews with the plant managers revealed that the relevant MACS attributes had existed in their present form before the initial FA adoption and had remained unchanged during the period in which FA adoptions had taken place. Showing that certain characteristics and uses of MACS precede FA adoption and remain constant during the adoption process is necessary (but not sufficient) to demonstrate causality. However, since the advice about MACS and FA in the current literature is unsubstantiated by empirical research, and considering the absence of any empirical evidence to the contrary, the GMC study does provide some useful insights for executives who wish to make their MACS more supportive of FA adoption. Senior executives who wish to promote FA adoption in their firms are advised to set budget targets for periods longer than one year and to report performance on those long-range objectives. Furthermore, senior managers should provide lower-level managers with resources for long-range planning. In reference to adoption (as opposed to utilization) of FA, the GMC study counsels incremental rather than radical reform of performance measurement and investment justification systems. Properly used, traditional MACS can support strategies of aggressive investment in advanced manufacturing technologies.

6. REFERENCES

Bernard, P. (1986). "Structured Project Methodology Provides Support For Informed Business Decisions", **Industrial Engineering, 18** (3), 52-57.

Bessant, J.R. (1982). "Influential Factors in Manufacturing Innovation", **Research Policy, 11**, 117-132.

Blank, L. (1985). "The Changing Scene Of Economic Analysis For The Evaluation Of Manufacturing System Design And Operation", **The Engineering Economist, 30** (3), 227-244.

Bower, J. L. (1970). **Managing the Resource Allocation Process**, Boston: Harvard University Graduate School of Business.

Brealey, R., S. Myers, G. Sick, and R. Whaley. (1986). **Principles of Corporate Finance.** Toronto: McGraw-Hill Ryerson Limited.

Bylinsky, G. (1986). "A Breakthrough In Automating The Assembly Line", **Fortune, 113** (11), 64-66.

Canada, J.R. (1986). "Annotated Bibliography On Justification Of Computer-Integrated Manufacturing Systems", **The Engineering Economist, 31** (2), 137-150.

Economic Council of Canada. (1987). **Making Technology Work: Innovation and Jobs in Canada.** Minister of Supply and Services Canada.

Finnie, J. (1988). "The Role of Financial Appraisal in Decisions to Acquire Advanced Manufacturing Technology", **Accounting and Business Research, 18** (70), 133-139.

Haka, S.F. (1987). "Capital Budgeting Techniques and Firm Specific Contingencies: A Correlational Analysis", **Accounting, Organizations, and Society, 12** (3), 366-376.

Hayes, R.H. and W. J. Abernathy. (1980). "Managing Our Way To Economic Decline", **Harvard Business Review, 58** (4), 67-77.

Hayes, R.H. and D. A. Garvin. (1982). "Managing As If Tomorrow Mattered", **Harvard Business Review, 60** (2), 70-78.

Hopwood, A.G. (1972). "An Empirical Study of the Role of Accounting Data in Performance Evaluation", **Journal of Accounting Research, 10**, 156-182.

Howell, R.A., J. D. Brown, S. R. Soucy, and A. H. Seed. (1987). **Management Accounting in the New Manufacturing Environment.** Montvale, N.J.: National Association of Accountants.

Huber, R.F., (1985). "Justification: Barrier To Competitive Manufacturing", **Production**, September, 46-51.

Jones, D.T. (1985). "The Internationalization of the Automobile Industry", **Journal of General Management, 10** (3), 23-44.

Kaplan, R.S. (1983). "Measuring Manufacturing Performance: A New Challenge For Managerial Accounting Research", **The Accounting Review, 59** (4), 686-705.

Kaplan, R.S. (1984). "The Evolution Of Management Accounting", **The Accounting Review, 59** (3), 390-418.

Kaplan, R.S. (1985). "Quantitative Models For Management Accounting In Today's Production Environment (Revised)", **Fourth Annual Deloitte, Haskins and Sells Accounting Research Symposium**, January, London Business School.

Lorsch, J.W. and J. J. Morse. (1974). **Organizations and Their Members**. New York: Harper & Row.

McLean T. (1986). "Manufacturing: Competitive Edge Or Corporate Millstone?", **The Accountant's Magazine**, May, 61.

Merchant, K. A. (1981). "The Design of the Corporate Budgeting Systems: Influences on Managerial Behavior and Performance", **The Accounting Review, 16** (4), 13-829.

Merchant, K. A. and W. J. Bruns, Jr. (1986). "Measurements To Cure Management Myopia", **Business Horizons, 29** (3), 56-64.

Pearson, G. (1986). "The Strategic Discount -- Protecting New Business Projects Against DCF", **Long Range Planning, 19** (1), 18-24.

Pinches, G.E. (1982). "Myopia, Capital Budgeting and Decision Making", **Financial Management, 11** (3), 6-19.

Schwarzbach, H.R. (1985). "The Impact Of Automation On Accounting For Indirect Costs", **Management Accounting, 58** (6), 45-50.

Sheridan, T. (1986). "How To Account For Manufacturing", **Management Today**, August, 61-74.

Woods, M., M. Pokorny, V. Lintner, and M. Blinkhorn. (1985). "Appraising Investment In New Technology: The Approach In Practice", **Management Accounting (U.K.), 63** (9), 42-43.

MANAGING THE FLEXIBILITY OF FLEXIBLE MANUFACTURING SYSTEMS FOR COMPETITIVE EDGE

Chen-Hua Chung
Department of Decision Science and Information Systems
College of Business and Economics
University of Kentucky
Lexington, KY 40503-0034

and

In-Jazz Chen
Department of Quantitative Business Administration
College of Business Administration
Cleveland State University
Cleveland, OH 44115

ABSTRACT

Flexible Manufacturing Systems (FMS) are believed to be a major means for improving both production flexibility and productivity. If well managed, FMS can be a formidable competitive weapon for manufacturing firms. However, there exists confusion concerning how to define the concept of flexibility. One result is the misconception that flexibility may cause the decline of productivity. This paper reviews the existing studies on flexibility and presents a "flexibility map" which clarifies the relationship between various flexibility concepts and measures. This map also provides a foundation for exploring the issue of how flexibility can contribute to the firm's competitiveness. We further point out the importance of a Total System Flexibility (TSF) concept which considers two important flexibility factors: quickness of response to a change and economic response to the change. A numerical example of routing flexibility is used to demonstrate how flexibility can enhance the competitive or strategic value of FMS. Indeed, total system flexibility can increase rather than reduce productivity, and therefore enhance the firm's competitiveness.

1. INTRODUCTION

The United States has encountered formidable competitive challenges in the international trade arena in recent years. Some researchers see the loss of manufacturing as a natural consequence of the economic evolution around the world and claim that we should pay attention to services and deemphasize manufacturing. However, other researchers feel that the loss of manufacturing capability will severely reduce the standard of living of the United States, and will impair the ability to conceive and develop new products and technologies (Bitran et al. 1988).

Furthermore, customers' requirements have become more diversified and product life cycles have become much shorter. The focus of international competition has changed from cost to quality, reliability, and the ability to respond quickly and accurately to customer needs. The latter is what is called flexibility.

The flexible manufacturing system (FMS), which is an important class of automated manufacturing, was developed in the early 1970's to provide manufacturing firms with the ability to cope with dynamic demand patterns and the capability to achieve cost savings, quality improvements, and productivity gains. A U.S. Department of Commerce (1987) study estimates that as of 1981, there were 4500 Computer Aided Design (CAD) installations in the U.S. The number of CAD installations is expected to reach an astonishing level of 190,000 installations by 1995, a forty-fold increase in just 14 years. This study also estimated that the world market for industrial automation, including CAD, Computer Aided Manufacturing (CAM), Computer Aided Engineering (CAE), Computer Integrated Manufacturing (CIM), and computer based systems for production planning and control, which was about 15 billion dollars in 1983, is expected to grow to a level of 65 to 70 billion dollars by 1989 (see also Economic Commission for Europe 1986). All these massive investments are linked in one way or another to the attempt to enhance manufacturing flexibility (Swamidass 1988). No doubt, flexibility has become an important competitive weapon. More and more FMS's are expected to be installed in years to come.

Flexibility is the raison d'etre of the flexible manufacturing system concept and practice. Unfortunately, despite the numerous attempts to describe "flexibility", or "manufacturing flexibility", the terms are not well understood. Swamidass (1988) points out three definitional problems hindering the understanding of flexibility. First, the scope of the flexibility-related terms overlap considerably. Second, some terms are aggregates of others. Third, identical terms used by various writers do not necessarily mean the same thing.

This paper intends to sort out the "mess" of flexibility concepts so that their roles in enhancing competitiveness can be better understood. We first review the types of manufacturing flexibility defined in the literature. Then, a "flexibility map" is developed to identify the relationship between the various flexibility concepts. With the flexibility map, the flexibility-competitiveness connection becomes clear. For any type of flexibility to contribute to competitiveness, both quick response to a change and economic response to the change should be considered. We therefore define a Total System Flexibility (TSF) concept to incorporate these two factors. To further illustrate how to accomplish competitiveness through flexibility, we use an example of routing flexibility to define the measurements of flexibility and its strategic value. A numerical example is provided.

2. A REVIEW OF FLEXIBILITY CONCEPTS

Due to the increasing recognition of the importance of flexibility in decision-making and planning, there have been quite a few attempts to define the term "flexibility". For example, Mandelbaum (1978) defined flexibility as "the ability to respond effectively to changing circumstances". Two kinds of flexibility were identified: action flexibility and state flexibility. Action flexibility is the capacity for taking new external action to meet new circumstances. State flexibility is the capacity to continue functioning effectively despite the circumstance changes. Most flexibility concepts defined in the FMS literature imply state flexibility. A discussion of several different approaches for categorizing the types of flexibility follows.

Buzacott (1982) identifies two types of flexibility in a manufacturing system, namely, job flexibility and machine flexibility. Job flexibility concerns the ability of the system to cope with changes in the jobs to be processed by the system. A machine flexibility, on the other hand, is the ability of the system to cope with changes and disturbances at the machines or work stations (e.g., machine/work station breakdowns).

Browne et al. (1984) define eight types of flexibility:

(1) machine flexibility: the ease of making the changes required to produce a given set of part types;

(2) process flexibility: the ability to produce a given set of part types, each possibly using different materials in several ways; Buzacott (1982) called this "job flexibility", which is related to the mix of jobs which the system can process;

(3) product flexibility: the ability to changeover to produce a new set of products economically and quickly;

(4) routing flexibility: the ability to handle breakdowns and to continue producing the given set of part types;

(5) volume flexibility: the ability to operate an FMS profitably at different production volumes;

(6) expansion flexibility: the capability of building a system and expanding it easily and modularly as needed;

(7) operation flexibility: the ability to interchange the ordering (sequencing) of several operations for each part type; even though some operations require certain precedence structure, others need not to be processed in partial order; and

(8) production flexibility: the universe of part types that the system can produce.

Yamashina et al. (1986) define three types of flexibility: (1) variant flexibility, which is related to the mix of products (parts) which an FMS can process; (2) volume flexibility, which is related to the ability to change production volumes without causing difficulties to an FMS; and (3) short product life flexibility, which is concerned with design changes or introduction of new products to the FMS due to short product life.

Kusiak (1986) proposed four types of flexibility and the

appropriate measures to evaluate them: (1) flexible manufacturing module (FMM) flexibility measured by the number of parts that can be processed on the FMM; (2) material handling system (MHS) flexibility concerned with the ability to handle different parts in a number of different routes; (3) computer system flexibility measured by its adaptability to the changing functions; (4) organizational flexibility which consists of: (a) job flexibility related to the mix of parts which the FMS can process; (b) scheduling flexibility measured by the number of routes along which a given job can be manufactured; (c) short-term flexibility measured by the changeover cost between known production programs; (d) long-term flexibility measured by the setup cost of the new production tasks due to changes in the production program.

Swamidass (1988) suggests using the manufacturing flexibility continuum to describe three types of flexibility: (1) high-volume/low-variety flexibility, to achieve low cost, high volume production of very few models of stable design on a common line with ease of new product introduction; (2) mid-volume/mid-variety flexibility, to achieve moderate variety, midvolume production of different configurations with various routings and in lot sizes as small as one; (3) low-volume/high-variety flexibility, to achieve custom design and manufacture of an infinite variety of very low volume products requiring the capability to adjust rapidly to frequent product design or redesign.

In summary, all these types of flexibility fall into two categories: machine level flexibility and system management level flexibility. Manufacturing flexibility can be achieved at either or both levels. Machine level flexibility, which is predominantly technology-based, is concerned with the internal features which provide flexibility. That is, machines, material handling, automatic tool changing, and so forth, are designed to be more versatile and more complex, and to contain more sophisticated controls. On the other hand, flexibility can also be obtained at the system level. However, this system management flexibility is a complex blend of technological, human, process, and manufacturing factors. It is achieved not only through machine/technology flexibility, but by efficiently moving jobs between machines through the use of the material handling system so that all necessary operations can be processed.

3. FLEXIBILITY AND COMPETITIVENESS

As shown in the previous section, the literature has suggested a variety of lists and categories of flexibility concepts. Indeed, in spite of the extensive literature on flexibility, there exists confusion concerning how to best define the concept of flexibility and to address several of the unresolved issues (Falkner, 1986).

3.1 Flexibility Map

To better understand how flexibility contributes to competitiveness, a "flexibility map" has been developed to depict the relationships among the various flexibility concepts appearing in the literature (Figure 1). The detailed, functional relationships between the flexibility terms and between flexibility and competitiveness are part of ongoing research. However, the map provides a valuable scheme for understanding flexibility and its potential contributions to a firm's competitiveness.

Since most of the flexibility terms have been reviewed in the previous section, a brief explanation follows. First of all, we divide flexibility into action flexibility and state flexibility according to Mandelbaum's scheme (Mandelbaum 1978). We further classify state flexibility into system management level flexibility and machine level flexibility. The system management branch contains organizational systems and the supporting computer and information systems, and so flexibility is contained within each of these. Organizational management activities usually involve a cycle of planning, organizing and control, which roughly corresponds to the Anthony taxonomy of strategic planning, tactical planning and operational control (Anthony 1965). In the context of FMS's, strategic planning mainly consists of product planning (i.e., design) and process planning. Thus, at this level we have

Figure 1. The Flexibility Map

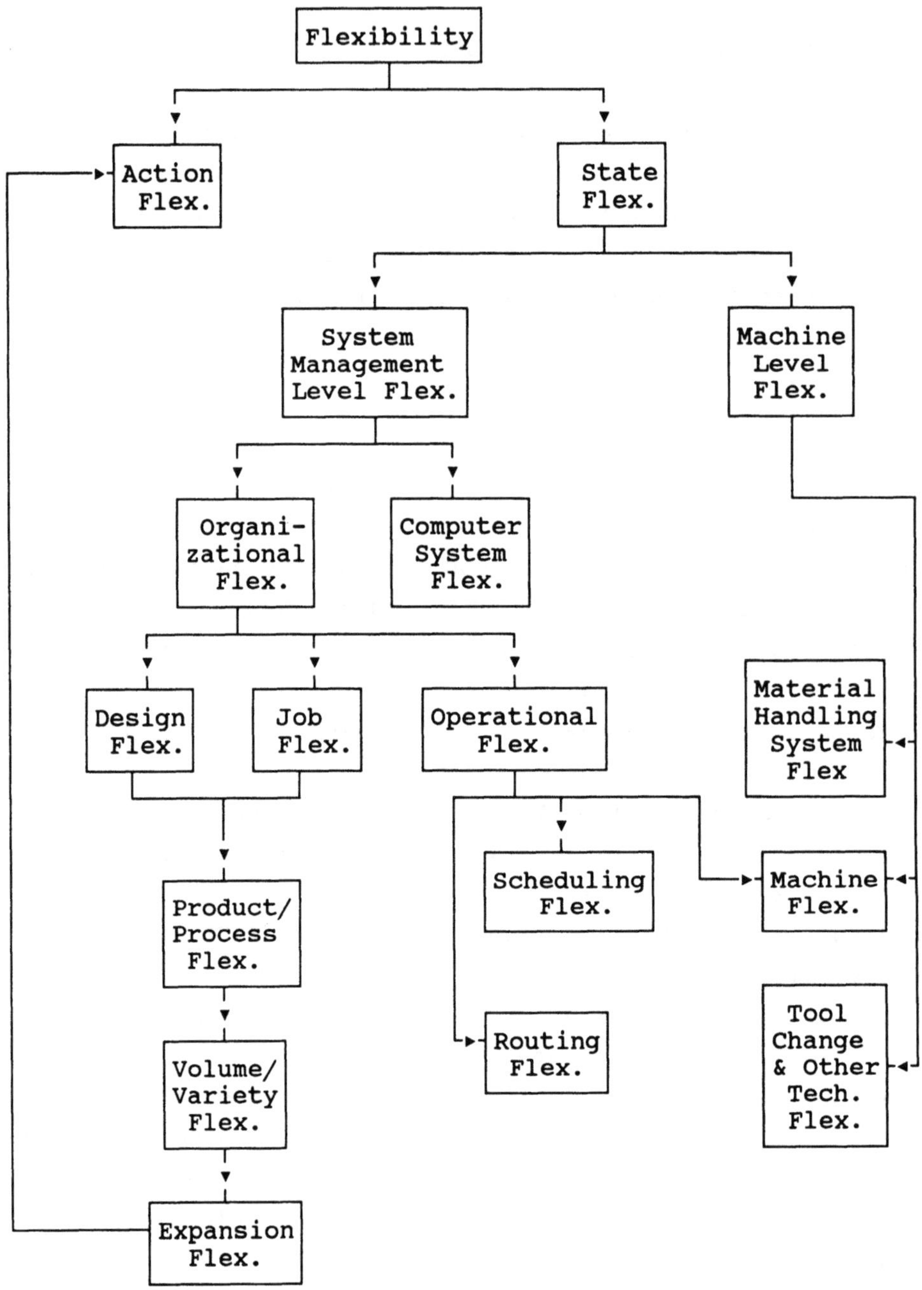

design flexibility which consists of product design flexibility and process design flexibility. We call the flexibility associated with organizing (or tactical planning) activities the job flexibility and further divide it into product flexibility and process flexibility. While the product and process flexibilities would be endowed by the product and process design activities, they are largely determined by other managerial decisions such as marketing or sales planning, production planning, part family grouping, and so forth. Thus, we place job flexibility (i.e., product and process flexibilities) at the tactical planning level. The volume/variety flexibility commonly addressed in the FMS literature (e.g., Yamashina et al., 1986, Swamidass 1988) actually results from both design flexibility (i.e., product and process design flexibilities) and job flexibility (i.e., product and process flexibility). The volume/variety flexibility also determines the firm's expansion flexibility. The latter, in turn, affects action flexibility. Operational flexibility consists of scheduling flexibility, routing flexibility and machine flexibility. The latter has been classified as the flexibility associated with the physical manufacturing system or the machine level flexibility. In addition to machine flexibility, the machine level flexibility also includes material handling system flexibility, automatic tool changing, and other technology-related flexibilities.

3.2 The Flexibility-Competitiveness Connection

Based upon the flexibility map shown in Figure 1, we can further explore how the various flexibilities can contribute to a firm's competitiveness. Such a flexibility-competitiveness connection is summarized in Figure 2. It should be pointed out that we omit computer system flexibility from the figure with the understanding that such flexibility will facilitate organizational management and therefore enhance competitive capability. Thus Figure 2 focuses upon other types of flexibilities and their contributions to the firm's competitiveness.

As mentioned earlier, action flexibility is the capability for

Figure 2. The Flexibility-Competitiveness Connection

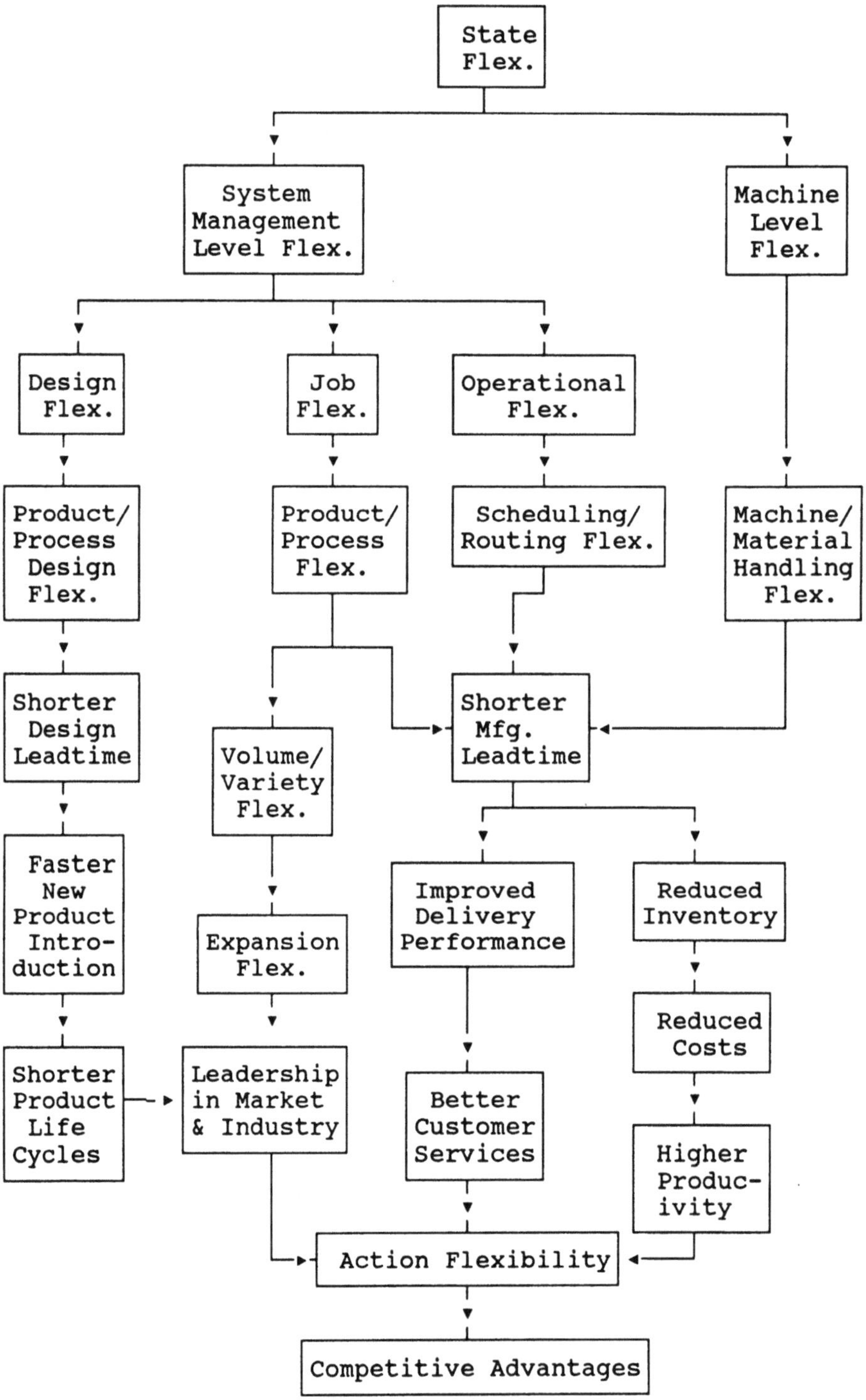

taking new external action to meet new circumstances. In today's turbulent environment, it is crucial for a firm to have such action flexibility to survive and to compete. McDougall and Noori (1986) point out that six major factors have led to a risky, uncertain environment. They are: short product life cycle, increased new product introduction, fragmented markets, technological change, unexpected competitors, and fluctuating demand. These factors create an extremely dynamic environment where firms face a great deal of uncertainty. Firms may use the flexibility derived from FMS as a weapon to react to change and gain long-term competitive advantage.

For example, a firm's design flexibility may help it shorten the design leadtime and therefore speed up and increase new product introduction. With design flexibility, the firm would be able to introduce new and improved products leading to shortened life cycles. The firm may thus gain competitive advantage by "leading" the market and the industry.

In addition to shortening design leadtime, it is also important to shorten manufacturing leadtime to improve delivery performance and reduce inventories. Lower inventory costs and the resulting decrease in overall costs imply higher productivity.[1] Better customer service and lower costs (and therefore, reciprocally, higher productivity) can certainly enhance the firm's competitiveness. Product and process flexibilities and machine flexibility are fundamental to the reduction of manufacturing leadtime. Routing flexibility also contributes to shrinking the manufacturing leadtime (see section 4).

The volume/variety flexibility, together with the resulting expansion flexibility can help the firm in implementing its positioning strategy. Hayes and Wheelright (1979) suggest that there is a close link between process and product life cycles. This linkage also provides a continuum of positioning strategies. Generally speaking, a process-focused strategy works well when a firm offers a wide variety of customized products. On the other

---

[1]Here productivity is defined as an output-to-input ratio. On the other hand, cost can be defined as the input-to-output ratio. Therefore, productivity and cost have a reciprocal relationship.

hand, a product-focused strategy makes sense in the environment of high-volume production of a few standard products with flow lines. Halfway between the two is an intermediate strategy. As shown in the manufacturing contingencies spectrum of Figure 3 (Jelinek and Goldhar 1983, Voss 1986), FMS and programmable systems (including CAD/CAM) belong to such an intermediate strategy. It is quite possible that the firm's action flexibility may move the position of FMS to the northeast corner of Figure 3. That is, FMS offers the capability of increasing both volume and variety of production.

Figure 3. Manufacturing Systems and Quantity – Product Variety Relationship

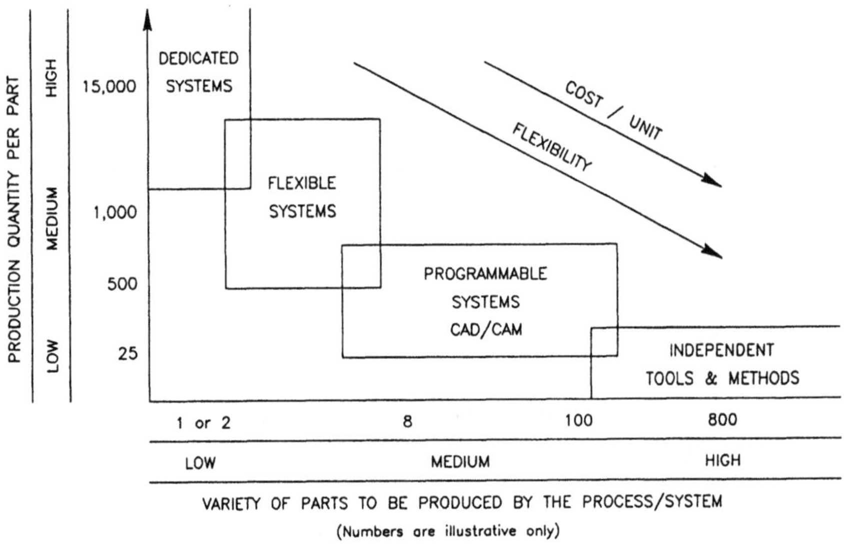

From the above analysis, it is obvious that state flexibility would enhance the firm's leadership position in market and industry, improving its customer service and increasing productivity. All these contributes to the firm's action flexibility and to its competitive advantage.

It is interesting to note that, with regard to the impact of flexibility on productivity, Buzacott (1982) and Buzacott and Mandelbaum (1985) found that most researchers who have considered flexibility seem to conjecture that the system performance declines as the flexibility increases. They further conjecture that those researchers view job flexibility as the ability of the system to cope with changes in the jobs to be processed. Zelenovic (1982), attempted to establish the relationship between flexibility and productivity. He discovered that the higher the degree of flexibility, the lower the level of productivity when the flexibility is achieved through production elements such as machines and work centers. Gustavsson (1984) also has pointed out that it is commonly believed that flexibility reduces system productivity.

This stream of research leads to an impression that flexibility reduces productivity, which is in conflict with the goal of implementing an FMS. We suspect that the decline in productivity may be largely due to the fact that manufacturing flexibility is pursued via production elements alone, such as machine tools (i.e. via job flexibility or machine flexibility). However, Figure 2 clearly shows that flexibility can enhance productivity. We further demonstrate in section 4 that routing flexibility (e.g., the use of an alternate routing policy) can, in fact, enhance system management flexibility and therefore increase rather than decrease productivity.

3.3 Total System Flexibility

With all the attention paid to flexibility issues, some authors seem to forget another key word in FMS -- the S for System. True flexibility should mean **total system flexibility** (TSF). An aggregate flexibility measure based upon the flexibility map of Figure 1 would provide an estimate for the TSF. Such a "flexibility accounting", although desirable, is complicated and yet to be developed. An alternate approach is to estimate a TSF for each flexibility term in Figure 1. In this paper, we only present a conceptual scheme for defining such TSF. Detailed flexibility accounting is part of on-going research. Other attempts to measure flexibility of FMS can be found in Chatterjee et al. (1984), Kumar (1986), Upton and Barash (1988), Hutchinson and Sinha (1989), Brill and Mandelbaum (1989), among others.

Masuyama (1983) suggests that FMS system flexibility can be captured by two factors, namely, (1) the quick response to a change (Q), and (2) the economical response to the change (E). Quickness can be evaluated as the leadtime between customer's order receipt and the completion of products. For a quick response to change, the leadtime needs to be minimized.

The system can be made capable of responding to a change quickly at the expense of enormous amount of capital investment. Therefore, quick response to a change alone is inadequate for the evaluation of the TSF. For example, an FMS using a large number of versatile machine tools, sophisticated industrial robots and material handling system might be capable of responding to a change very quickly. However, the extensive capital investment may not be economically justifiable. Factors such as inventory level and machine utilization should be evaluated.

The TSF defined below captures both the quickness (Q) and the economical (E) response to a change. Depending on the situation, different weights (α) can be assigned to these two goals in the measurement of TSF.

Definition 3.3.1: The total system flexibility (TSF) of an FMS,

$$\mathrm{TSF} = \alpha\, Q + (1 - \alpha)\, E$$

where $0 \leq \alpha \leq 1$

Definition 3.3.1 provides a conceptual framework for understanding and evaluating the TSF of an FMS. It allows situations where quick response to a change and economic response to a change may be conflicting goals. For example, sometimes firms have to determine whether to maximize the customer service level or to minimize the cost, or to optimize one of the goals while keeping the other at a prespecified level.

In Definition 3.3.1, TSF was represented as a linear combination of Q and E. Certainly, other types of relationships between Q and E such as a nonlinear Q-E frontier are possible. This is illustrated in the following definition.

Definition 3.3.2: The total system flexibility of an FMS,

$$\mathrm{TSF} = Q^{\alpha} E^{\beta}, \quad \alpha + \beta = c,$$

where $c \geq 1$ is a constant.

Definition 3.3.2 has an advantage over Definition 3.3.1 in

that it does not exclude the possibility for both flexibility and productivity to increase simultaneously. That is, both Q and E can increase at the same time even when bounds are placed on the weights assigned to each factor.

Definitions 3.3.1 and 3.3.2 only offer a conceptual construct for showing the nature and the importance of both Q and E factors. It would be difficult for one to measure the TSF using the above formulas. Particularly, the assignment of weights to each factor may be subjective (only one of the two weights needs to be determined because the two weights are complementary). Also, the procedure for measuring the Q and E may vary according to the different flexibility terms in Figure 1. It is more plausible to develop a surrogate measure for TSF. This will be demonstrated via examples in the next section.

4. A ROUTING FLEXIBILITY EXAMPLE

In this section, we use a routing flexibility example to demonstrate how flexibility can contribute to productivity improvement and therefore to the enhancement of a firm's competitiveness. The routing decision in an FMS is concerned with arranging the jobs through a set of machines that are capable of processing the required operations with the objective of either maximizing the resource utilization, or minimizing the job lateness, or both. However, the assignment of operations to machines is a loading problem. Thus, the loading decisions should be made prior to the routing decisions. There are various possible loading objectives (Stecke 1983). In the spirit of maximizing the flexibility of FMS, we assume that the loading task is to maximize the number of ways operations can be assigned to machines subject to the limited resource constraints such as pallets, fixtures, and tool magazine capacity. This results in the maximal number of alternate routes. For detailed discussions on the loading and the routing models and the interfaces between the two problems, see Chen (1989).

4.1 Routing Flexibility

Chatterjee et al. (1984) define routing flexibility for each **individual** part as the cardinality of the set of routing for that part. Yao (1984) describes an entropic measure of routing flexibility. He considers the flexibility to be a function of the breakdown frequency of the machines in the system. Here we define routing flexibility for the **whole** system as follows:

Definition 4.1.1: The routing flexibility of an FMS is defined as follows:

$$RF = \frac{\sum_{h=1}^{H} \rho_h}{H}$$

where

H: total number of parts

ρ_h: the number of feasible routes that part type h can flow through the system.

The routing flexibility defined above gives the average number of alternate routes for processing each part type. It should be noted that if the average number of alternate routes for each part is m (i.e., RF=m) and there are n part types, then the total number of possible **combinations** of alternate routes for the system, on the average, would be m^n. However, the actual total number of route mixes would be m x n.

Sometimes what is more important is the number of feasible routes which constitute the optimal routing mix, rather than the total number of feasible routes an FMS can possess. In other words, those routes that are not used to achieve the optimal route mix may make little or no contribution to the system performance. In such a case, we can revise the above definition as follows :

Definition 4.1.2: The routing flexibility of an FMS is defined as follows:

$$RF' = \frac{\sum_{h=1}^{H} \lambda_h}{\sum_{h=1}^{H} \rho_h}$$

where

λ_h: the number of feasible routes which constitute the optimal routing mix in terms of routing objectives such as minimizing makespan.

This routing flexibility represents the percentage of the feasible route mixes which are needed to achieve the optimal routing performance. Thus, those route mixes which do not provide optimal routing decisions can be discarded. The resources (e.g., tools) that were originally assigned to these mixes by the loading decisions can be removed or reallocated. Cost savings would then be possible. The managerial implications are obvious.

The strategic value of routing flexibility can be evaluated in terms of total job completion time (i.e. job makespan). Let T_{FR} be the total job completion time required if a fixed route is used, and T_{AR} be the total job completion time needed if alternate routes are allowed. The strategic and/or competitive value of the routing flexibility then can be defined as follows:

Definition 4.1.3: The strategic/competitive value of routing flexibility is defined as:

$$V_{RF} = \frac{T_{FR} - T_{AR}}{T_{FR}}$$

The above definition clearly indicates the percentage reduction of total job completion time that routing flexibility can provide. It can be easily transformed into a dollar value, if desired. It is interesting to note that Mandelbaum and Buzacott (1986), in studying decision process flexibility, suggested that the worth of flexibility is not a measure of flexibility itself but a consequence of it. Definition 4.1.3 provides an example of their view of flexibility.

Bitran et al. (1988) pointed out that, in the new competitive environment, a firm's advantage often lies in its ability to reduce cycle times and minimize the length of time that a product remains in the shop, while reducing inventories and yet responding rapidly and accurately to customer needs.

Minimizing total job makespan will not only minimize the job processing time and waiting time (i.e. the length of time a product remains in the shop) but also will reduce inventories and increase the resource utilization. Therefore, if one can increase V_{RF} one can respond to a market change **quickly and economically** and thus achieve the competitive advantage. Indeed, Definition 4.1.3 provides an integrated measure for the competitive/strategic value of routing flexibility as well as a TSF measure for routing flexibility.

The above reasoning can be best illustrated by the Honda-Yamaha motorcycle "war" in the early 1980's when Honda's number one market share position was threatened by Yamaha. To counter the threat, Honda introduced 81 new models and discontinued 32 others for a total of 113 changes in its product line within eighteen months. On the contrary, without the system flexibility, Yamaha retired only 3 models and introduced 34 new models for a total of 37 changes. By January 1983, about a year after Honda's quick response counter attack, President Koske of Yamaha admitted defeat and ended the Honda-Yamaha war.

In summary, we have defined two measures for evaluating routing flexibility as well as the competitive/strategic value of the routing flexibility. Routing flexibility, as defined in 4.1.1, is readily obtainable since only the output from the loading decision is required. Therefore, it is a convenient estimate of how flexible an FMS is in directing parts through the system. This information is particularly useful when machine or tool failures occur.

Both Definitions 4.1.2 and 4.1.3 require solving the routing problem to find either the number of feasible routes which constitute the optimal solution or the makespan. Although a little more complicated, they provide insights into the benefit of the alternate routing capability. Definition 4.1.2 gives the ratio of the number of feasible routes which constitute the optimal routing mix to the total number of feasible routes. This ratio provides valuable information about the utilization of alternate routes and, hence, can be fed back to improve the loading decisions. For instance, a low ratio indicates that many of the feasible routes generated by the loading decision are not fully utilized. Resources such as tools which are allocated to generate those "useless" routes could be removed or reallocated without affecting the optimal system performance. Savings resulted from such reallocation would certainly help improve productivity. Furthermore, Definition 4.1.3 can be used to evaluate the competitive value of alternate routing in terms of total job makespan reduction.

4.2 A Numerical Example

The following numerical example demonstrates the use of flexibility measures and the competitive value of routing flexibility defined above, and clarifies the relationship between flexibility and productivity. The problem refers to an FMS with different Numerical Control (NC) machines which can perform the same set of operation(s) using the same set of tools but with different processing times (i.e., taking into account the tool efficiency) (Chen 1988). Suppose that the FMS consists of three NC machines, M_1, M_2, and M_3, which can process two types of parts, J_1 and J_2. The operation sequence for J_1 and J_2 and the corresponding machines required are given in Table 1, where the processing times required (in minutes) are given in the parentheses.

The information shown in Table 1 is the output from an FMS loading model. That is, part type J_1 requires two operations while part type J_2 requires three. The first operation of both part types J_1 and J_2 can be processed on either machine M_1 or M_2, and the second operation of both part types can be performed on machine M_2. The third operation of part type J_2 can be processed on either machine M_2 or M_3. Consequently, part type J_1 has two routing alternatives whereas part type J_2 has four routing alternatives to

Table 1. The Loading Decision and Processing Times

	Operation #1	Operation #2	Operation #3
J_1	M_1 (9) M_2 (6)	M_2 (5)	-
J_2	M_1 (9) M_2 (6)	M_2 (5)	M_2 (23) M_3 (18)

flow through the FMS. The routing alternatives are listed in Table 2.

Table 2. The Routing Alternatives for Part Types

	Route Alternatives	Route Description
J_1	A	M_1 - M_2
	B	M_2 - M_2
J_2	C	M_1 - M_2 - M_2
	D	M_1 - M_2 - M_3
	E	M_2 - M_2 - M_2
	F	M_2 - M_2 - M_3

Typical loading models which generate the decisions shown in Table 1 can be found in Chen (1988). The output obtained from the loading models is a set of routing alternatives for each part type. However, which route or route mix should be used for each part type in order to optimize system performance is determined by the routing model which, based on the loading decisions, attempts to optimize the system performance such as total job makespan. The routing model used to solve the above example is as follows:

4.3 The Routing Model

Minimize MS

s.t.

$$\sum_{h=1}^{H} \sum_{k \in K} \tau_{hjk} \cdot n_{hk} \leq MS, \quad j = 1, 2, \ldots, N \tag{1}$$

$$\sum_{k \in K} n_{hk} \geq D_h, \quad h = 1, 2, \ldots, H \tag{2}$$

$$n_{hk} \geq 0, \quad \text{for all } h, k \tag{3}$$

where the subscripts and parameters are defined as follows:

subscripts:

operation $i = 1,2,\ldots,M$

machine $j = 1,2,\ldots,N$

routes $k = 1,2,\ldots,K$

part type $h = 1,2,\ldots,H$

parameters:

MS: total job makespan

D_h: demand for part type h

n_{hk}: the number of part type h processed via route k

τ_{hjk}: time to perform the operation(s) of part type h on machine j if route k is used

Constraint (1) establishes the relationship between the makespan and the total machine work time on each machine. This ensures that the job's makespan should be greater than or equal to the total work time on each of the machines. Constraint (2) guarantees that the demand for each part type is met. Finally, the number of each part type produced using a specific route should be as stated by nonnegative constraint (3). The solution to the routing model is summarized in Table 3.

As mentioned earlier, one of the advantages of using Definition 4.1.1 is that it is readily obtainable. That is, only

Table 3. Routing Decision and Total Job Makespan Under Alternate Routing and Fixed Routing

Part Types	Alternate Routing						Fixed Routing					
	% of Parts Processed Via Route						% of Parts Processed Via Route					
	A	B	C	D	E	F	A	B	C	D	E	F
20 units of J_1	100%	–	N/A	N/A	N/A	N/A	100%	N/A	N/A	N/A	N/A	N/A
20 units of J_2	N/A	N/A	15%	55%	–	30%	N/A	N/A	N/A	100%	N/A	N/A
Total Job Makespan	MS = 306 minutes						MS = 660 minutes					
Completion Time Reduction	$\frac{660 - 306}{660} = 53\ \%$											

the output from the loading model is needed. For example, Table 2 shows the routing flexibility RF = (2 + 4)/2 = 3. That is, on average, each of part types J_1 and J_2 will have about three alternative routes. Consequently, on the average, there are 9 (i.e., 3 x 3) routing alternatives for the FMS to process part types J_1 and J_2. It should be noted that the actual number of routing alternatives is 8 (i.e., 2 x 4). However, Definition 4.1.1 does provide an indicator of how flexible the FMS is in directing parts through system.

On the other hand, the routing decision is required when using Definition 4.1.2. As shown in Table 3, the routing flexibility RF' (Definition 4.1.2) of the FMS is (1 x 3)/(2 x 4) = 0.375. This figure indicates that approximately 38% of the feasible route mixes generated by the loading model are needed to achieve the optimum system performance -- minimizing job makespan. It is clear that 62% of the feasible route mix alternatives are not utilized. Consequently, tools originally assigned to these route mixes by the loading model can be removed or reallocated without affecting system performance.

As far as the competitive value of alternate routing capability is concerned, Table 3 shows that the total job makespan can be reduced by 254 minutes or 53% if the routing flexibility is utilized to enhance system management level flexibility. The job completion time reduction also implies work-in-process inventory reduction, cost reduction, and productivity improvement. The competitive and strategic value of such changes is obvious. More importantly, the job completion time reduction improves due date performance and therefore customer satisfaction. Faster delivery of products to customers (i.e., the quickness factor) and reduced inventory and overall costs (i.e., the economic factor) would certainly give the firm some competitive edge.

Furthermore, the value of alternate routings is even more significant in the case of machine breakdown. For example, suppose that machine M_3 goes down or a defective part is detected from machine M_3. With alternate routing capability, machine M_3 can be automatically bypassed. That is, the scheduler can simply redirect part type J_2 through the FMS using routes C and/or E so that the detrimental effects of the disruption can be minimized.

As mentioned in Section 3, the factor of quick response to a change, Q, is the leadtime between the receipt of a customer's order and the completion of that order. Since manufacturing leadtime is determined by job processing time and job waiting time, one can reduce the manufacturing leadtime by reducing the job completion time. The latter also results in inventory reduction. The economic implication is obvious. Thus, one can enhance both the quick response factor and the economic factor, and, therefore, the total system flexibility (TSF).

Table 3 shows that the job completion time can be reduced by 53% if an alternate routing policy is adopted. Thus, alternate routing increases the total system flexibility by enhancing the quick response factor. On the other hand, productivity had also increased since the input, in term of machine time, has been reduced. Therefore, it can be concluded that the increased system flexibility which resulted from routing flexibility can also improve productivity. This is another economic benefit brought about by increasing system flexibility.

The alternate routing strategy usually requires more tools when compared with the use of a fixed routing for a given FMS. In situations where tools wear out easily and need to be replaced quite often, the required resources for alternate routings may increase, and possibly hamper the productivity performance. Further economic analysis would then be needed.

5. CONCLUSION

The Flexible Manufacturing System (FMS) can improve both production flexibility and productivity by responding quickly and accurately to customer needs. Thus, FMS can become a competitive weapon for manufacturing firms. However, in spite of an extensive literature, there exists confusion concerning how to define the concept of flexibility. This may have led to some misconceptions about flexibility. For example, it is commonly believed that higher degrees of flexibility would lower the level of productivity. This is certainly not the purpose of FMS.

In this paper, we first reviewed the existing categorization of the flexibility concepts and developed a "flexibility map" to clarify the relationships between the various flexibility terms. Based on the flexibility map, we further explored the flexibility - competitiveness connection. It was conjectured that the state flexibility of a manufacturing system can enhance the firm's capability of increasing productivity, improving customer services, and becoming the market and industry leader. These capabilities provide action flexibility which helps the firm to gain competitive edge. We then presented a conceptual scheme of Total System Flexibility (TSF) which considers both quickness to response to a change and the economic response to the change. An example of routing flexibility was then used to demonstrate how flexibility can contribute to competitiveness. Basically, routing flexibility can help minimize job makespan which results in faster delivery of products to customers (i.e., the quickness factor) and reduced inventory and overall costs (i.e., the economic factor). These benefits would certainly give the firm some competitive edge in the market place. Thus, we have demonstrated that routing flexibility can enhance both the total system flexibility and provide competitive and strategic value. Moreover, total system flexibility can increase rather than decrease productivity, if it is enhanced via routing flexibility.

Swamidass (1988) cautioned that technology often comes to a manager's mind when considering manufacturing flexibility for competitive advantage. While it may be very tempting to do so, it may turn out to be a serious mistake to consider manufacturing flexibility only in terms of process technology or hardware. Machine level flexibility alone (e.g. versatile NC machines) would not seem to be adequate in assuring the competitive edge. It can elevate competitiveness only with the help of the flexibility in system management (e.g., an alternate routing policy).

In addition to further study on flexibility accounting as mentioned earlier, more research is required to develop situation-specific flexibility policies. That is, we require knowledge concerning the type of flexibilities which should be emphasized under different environmental changes. The total system flexibility concept is another important issue that needs further exploration. Although the scheme presented in this paper provides a good starting point, it is necessary to investigate the more specific relationships between the quickness factor and the economic factor, particularly in the context of the flexibility map.

6. REFERENCES

Anthony, R. R. (1965). "Planning and Control Systems: A Framework for Analysis," **Harvard University Graduate School of Business Administration Studies on Management Control.**

Brill, D. and M. Mandelbaum. (1989). "On Measures of Flexibility in Manufacturing Systems," **International Journal of Production Research, 27**, 747-756.

Bitran, G. R., W. Skinner, and W. I. Zangwill. (1988). "Operation Research in Manufacturing," **Operations Research, 36** (2), 172.

Browne, J., D. Dubois, K. Rathmill, S. Sethi, and K. E. Stecke. (1984). "Classification of Flexible Manufacturing Systems," **The FMS Magazine**, 114-117.

Buzacott, J. A. (1982). "The Fundamental Principles of Flexibility in Manufacturing Systems," **Proc. of 1st International Conf. on FMS**, October, 13-22.

Buzacott, J. A. and M. Mandelbaum. (1985). "Flexibility and

Productivity in Manufacturing Systems," **Proceedings of IIE Conference**, Chicago, 890-905.

Chaterjee, A., M. A. Cohen, W. L. Maxwell, L. W. and Miller. (1984). "Manufacturing Flexibility Models and Measurement", **Proceedings of the First ORSA/TIMS Conference on FMS**, Ann Arbor, Michigan.

Chen, I. J. (1988). "Loading and Routing: Two Functions of Planning and Control in Flexible Manufacturing Systems," working paper, Department of Decision Sciences and Information Systems, University of Kentucky.

Chen, I. J. (1989). "Interfacing The Loading and Routing Decisions for Flexibility and Productivity of Flexible Manufacturing Systems," Unpublished Doctoral Dissertation, University of Kentucky.

Economic Commission for Europe. (1986). **Recent Trends in Flexible Manufacturing System Industry**, New York: United Nations.

Falkner, C. H. (1986). "Flexibility in Manufacturing Plants", **Proceedings of The Second ORSA/TIMS Conference on Flexible Manufacturing Systems**, Ann Arbor, Michigan, 95-106.

Gustavsson, S. O. (1984). "Flexibility and Productivity in Complex Production Processes," **International Journal of Production Research, 22** (5), 801-808.

Hayes, R. H. and S. C. Wheelright. (1979). "Link Manufacturing Processes and Product Life Cycles," **Harvard Business Review**, January-February, 133-140.

Hutchinson, G. K. and D. Sinha. (1989). "A Quantification of the Value of Flexibility," **Journal of Manufacturing Systems, 8** (1), 47-57.

Jelinek, M. and J. D. Goldhar. (1983). "The Interface Between Strategy and Manufacturing Technology," **Columbia Journal of World Business**, Spring, 16-36.

Kumar, V. (1986). "On Measurement of Flexibility in Flexible Manufacturing Systems: An Information-Theoretic Approach," **Proc. of the 2nd ORSA/TIMS Special Conference on FMS**, Ann Arbor, Michigan, August, 131-143.

Kusiak, A. (1986). "Parts and Tools Handling Systems," **Modeling and Design of Flexible Manufacturing Systems**, Amsterdam: Elsevier Science Publishers, 99-109.

Mandelbaum, M. (1978). "Flexibility in Decision Making: An Exploration and Unification," Ph.D. Thesis, Dept. of Industrial Engineering, University of Toronto, Toronto, Ont.

Mandelbaum, M. and J. A. Buzacott. (1986). "Flexibility and Its Use: A Formal Decision Process and Manufacturing View," **Proc. of the 2nd ORSA/TIMS Special Conference on FMS**, Ann Arbor, Michigan,

August, 119-130.

Masuyama, A. (1983). "Idea and Practice of Flexible Manufacturing System of Toyota," **Proceedings of 7th International Conference on Production Research**, Windsor, Ontario, 584-590.

McDougall, G. H. G. and H. A. Noori. (1986). "Manufacturing-Marketing Strategic Interface: The Impact of Flexible Manufacturing Systems," **Modeling and Design of Flexible Manufacturing Systems**, Elsevier Science Publishers.

Stecke, K. E. (1983). "Formulation and Solution of Nonlinear Integer Production Planning Problems for Flexible Manufacturing Systems," **Management Science, 29** (3), 273-288.

Swamidass, P. M. (1988). **Manufacturing Flexibility, Monograph No. 2**, Operation Management Association, January.

Upton, D. M. and M. M. Barash. (1988). "A Grammatical Approach to Routing Flexibility in Large Manufacturing Systems," **Journal of Manufacturing Systems, 7** (3), 209-221.

U.S. Department of Commerce. (1987). **A Competitive Assessment of the U.S. Flexible Manufacturing System Industry**, July.

Voss, C. A. (1986). **Managing New Manufacturing Technologies**, Monograph Series, No. 1, Operations Management Association.

Yamashina, H., K. Okamura, and K. Mutsomoto. (1986). "Flexible Manufacturing Systems in Japan: An Overview," **Proceedings of the Fifth International Conference on Flexible Manufacturing Systems**, November, 405-416.

Yao, D. D. (1984). "An FMS Network Model with State-Dependent Routing," **Proc. of the First ORSA/TIMS Conf. on FMS**, Ann Arbor, Michigan.

Zelenovic, D. M. (1982). "Flexibility - A Condition for Effective Production Systems," **International Journal of Production Research, 20** (3), 319-337.

ON AN ECONOMIC MEASURE OF SUPER-FLEXIBILITY IN AN UNCERTAIN ENVIRONMENT

Abraham Mehrez
Department of Industrial Engineering and Management
Ben-Gurion University,
Beer-Sheva, Israel 84105

ABSTRACT

A probabilistic economic measure is suggested to assist in the evaluation of the flexibility of Flexible Manufacturing Systems (FMS's). After deriving its analytical properties, the manner in which this measure could assist the decision-maker in the FMS project selection problem is shown. The measure determines the willingness of a risk-neutral decision maker to pay for a so-called "super flexible" joint technology, and considers the flexibilities of both manufacturing systems being evaluated instead of just merely selecting the better of the two. The measure finds a maximum price that a risk-neutral decision maker would pay, given that the economic value of a manufacturing system is evaluated via a stochastic one-period model.

The analysis of the statistical properties of the measure is based on the evaluation of the value of information for a risk-neutral decision maker facing a project-selection problem. There are several advantages to using the proposed measure. First of all, it is evaluated in monetary terms. Secondly, an upper bound on the maximum price that a risk-neutral decision-maker would pay can be provided based on very limited information about the underlying distributions that describe the economic values of each manufacturing system. Thirdly, it may guide the decision-maker to look at possible marginal improvements of flexibility through a technological change. The limitations of the suggested measure are also discussed, with particular reference to future research focused on the important engineering implications of the approach suggested here.

1. INTRODUCTION

Increasingly, as flexible manufacturing technologies become available across a broad range of applications (see for example Jaikumar, 1986), more and more firms must make decisions about the adoption of Flexible Manufacturing Systems (FMS's). So far, few theoretical and empirical economic justifications for FMS's have been discussed in the literature (see Snader, 1986; Canada, 1986) concerning the savings attributable to FMS's. These inadequacies seem due in part to the difficulties of explicitly quantifying the benefits from investment in "flexibility."

In general terms, "flexibility" is viewed as the ability of a system or decision-making process to effectively respond to change. This concept has been examined by several authors spanning several disciplines (Cunningham and Mandelbaum, 1985; Rosenhead and Gupta, 1968; Harrigan, 1985; Merkhofer, 1977; Marshak and Nelson, 1986; Pye, 1978; Rosenhead, et al., 1972; Brill and Mandelbaum, 1989; Mandelbaum and Buzacott, 1986).

The need for a common treatment has recently become more pressing as the strategic advantages of manufacturing flexibility have focused managerial attention on FMS (Hayes, Wheelwright and Clark, 1988). This subject is drawing research attention from several disciplines. For example, Stecke, et al. (1985) discuss **machining** flexibility and **assembly system** flexibility; Buzacott (1982) discusses **action** flexibility, **state** flexibility, **job** flexibility and **machine** flexibility; Zelenovic (1982) addresses **strategic planning** flexibility, **design** flexibility and **adaptation** flexibility; and Yao (1985) focuses on **routing** flexibility. Kumar (1986) also recognizes that there can be **loading** flexibility, **material handling** flexibility, **information flow** flexibility and so on.

In what follows, we review the literature in terms of current approaches and problems to be addressed in the measurement of flexibility. Next, we provide some motivation for the proposed economic measure, including an illustrative example. Finally, the measure, along with an analysis of its properties and possible applications, is described.

2. LITERATURE REVIEW

Current approaches to measuring flexibility include the information-theoretical approach (Kumar, 1986; Kapur, et al., 1985); the Petri net approach (Barad and Sipper, 1988); and the robustness approach (Rosenhead and Gupta, 1968). Their suggestions are applicable in particular contexts. For example, the entropy H-measure has been proposed as a means to capture loading and operations flexibility (Kumar, 1986, 1987) or routing flexibility (Yao, 1985). The entropy function is a mechanistic measure that has been used successfully to represent degrees of concentration or dispersion in a variety of applications. However, it is best suited for static or at least stationary phenomena.

On the other hand, the "robustness" approach (Rosenhead and Gupta, 1968; Rosenhead, et al., 1972), which is used to express flexibility in terms of uncertain possibilities over an extended period of time, may not be suitable for measuring the short-term flexibility that reflects and correlates with operational efficiency. Finally, Brill and Mandelbaum (1989) develop a measure-theoretical approach and provide measures of flexibility relative to a background set of possible tasks to be completed. The measures incorporate weights of task importance and machine-task efficiency ratings.

From a theoretical perspective, two separate sources of difficulty have to be overcome for measures of flexibility to be made operational. The first problem generally encountered stems from the unavailability of sufficient forecast data. The notion of flexibility entails gazing at the future, or alternative futures, to account for the possibility of a change in direction. Not only is the future hard to predict, but it is difficult to specify a subjective distribution function that reflects a decision-maker's uncertainty about future events requiring evaluation (Ackoff, 1970, 1981). For most problems, forecasting errors or lack of sufficient data over an infinite horizon may result in suboptimal decisions, as discussed by Lee and Denardo (1986) and Sethi and Chand (1979).

A second problem arises from difficulties in deriving aggregate utilities from individual or group preferences. As a case in point, the decomposition of a multi-attribute von Neumann-Morgenstern or non-von Neumann-Morgenstern utility function for a set of behavioral assumptions or axioms is still not accomplished, even when only a single decision-maker is involved. This problem has been well outlined in the decision theory literature (Arrow, 1951; Demski, 1980; Keeney and Raiffa, 1976; Bell and Farquhar, 1986) but still awaits an acceptable solution. Additionally, subjective probabilities of future events and their desirabilities are hard to encode: their evaluation and aggregation are cumbersome and error-prone, even when theoretically justifiable (see for example, Wallsten and Budesco, 1981). As a result, a variety of non-utility based approaches are currently being evaluated by researchers. These include the Analytic Hierarchy Process (Saaty, 1980), multi-objective models (Rosenthal, 1985), goal programming (Ignizio, 1976), and multi-criterion models (Lazing, 1986), among others.

From a practical perspective, evaluation of "flexibility" requires the solution of some basic problems of uncertainty. The **design** problem can be decomposed into: product design, part design, process planning, system design, equipment selection and layout design (Kusiak, 1987). On the other hand, the FMS **operational** problem may have the following structure: aggregate planning, resource grouping, disaggregate planning and scheduling (Kusiak, 1987), (see Buzacott and Yao, 1986; Stecke, 1985; Kusiak, 1986, 1987; Suri, 1985). The dynamic nature of the manufacturing environment, emanating from highly variable factors, such as customer orders, material and capacity constraints and machine or work station availability, makes it difficult to obtain global solutions for the just noted inter-related problems. This observation applies to any type of manufacturing system with or without advanced technology. Thus, it is a difficult task to measure accurately the long-run performance of a system based on the above-mentioned problems (Stecke, 1985). The traditional multi-attribute utility approach (see for example Groover and Zimmers, 1984; Stecke, 1985; Sink and Devires, 1984), which considers measures of cost, time, quality and reliability, as well as other factors affecting the capability and the flexibility of a manufacturing system, does not propose a unique economic measure for evaluating and comparing the flexibilities of advanced manufacturing technologies.

3. BACKGROUND ON THE APPROACH

Given these theoretical and practical difficulties, an economic measure for evaluating flexibility is proposed. The recent literature has established the importance of taking a step back and looking at the big picture, namely the choice of the technology within which engineers choose to operate (Abernathy, 1978; Kochan, 1986; Nord and Tucker, 1987). Top executives have a choice between a stable strategy of commitment to a specific technological family or thrust, and a shift to a newer technology. Effective top-level steering requires decisions to be made about research and development (R&D) spending, as well as the degree to which computer-aided design (CAD) and computer-aided manufacturing (CAM) should be introduced. Even if the CAD/CAM issue is not confronted outright, the management team has to take a long-term view of its plan for development.

Recently, Monahan and Smunt (1989) divided the problems of acquiring automation into two groups. The first deals with the general problem of the adoption of new technology. The second focuses on the particular problem of converting to new production processes. The complexity of jointly determining the production technology and the mix, scope and volume of products that can or will be produced, calls for a measure that helps the decision-maker to evaluate and compare different technologies or manufacturing systems. We determine a measure based on the Mehrez, et al. (1987) model, which evaluates the economic value of a manufacturing system via a stochastic one-period model. In light of the two separate sources of difficulty that have been mentioned before with regard to operational measures of flexibility (i.e., lack of forecast data and determining aggregate preferences), and to simplify the analysis, we place some restrictions--namely that technological flexibility is evaluated given that **only two** manufacturing systems are considered.

We assume a variety of economic states of the environment,

each of which requires a different operational solution. Flexibility is represented by the willingness of a risk-neutral decision maker to pay for a technologically feasible alternative that provides the flexibilities of both manufacturing systems. This alternative is called "super flexible" joint technology, and may not exist in reality. Moreover, it may not fit into the "exact" problem of a particular decision-maker. However, it provides economic information which may induce the decision-maker to search for possible marginal flexibility improvements via technological change. The analysis is based on the evaluation of the value of information for a risk-neutral decision maker facing a project-selection problem (see Mehrez and Stulman, 1982; Fatti, et al., 1987).

4. MOTIVATING EXAMPLE

Recently, Mehrez, et al. (1987) reported about a consulting study dealing with a firm that is considering the purchase of an FMS to replace an existing operation. Two basic conclusions can be drawn from this study. First of all, the annual cost and revenue depend primarily on the state of the economy, including the demand for existing and new products. Secondly, various alternatives exist for an FMS, depending on the system's equipment, configuration and management approach. Thus, the decision on which system to install should take into account its "flexibility" to operate under different conditions relative to other systems. To illustrate this point, we describe only two alternatives: the existing (non-FMS) system, and an FMS system.

Each alternative must be capable of producing four different products, which are gear reduction units. This particular operation assembles a housing, bearings, gears springs and fasteners into the final product, and performs certain quality and product performance checks. The existing system requires four operators and uses two small, old presses. The proposed FMS consists of an automatic material handling system, a multi-station indexing turntable, two robots, two special presses and six automatic feeders, and requires one operator. Clearly, other FMS's with different types of equipment could be considered.

The expected total annual costs for the existing system (CE) depends on the operating and maintenance cost of an operator via the following formula:

$$C_E = \left(\frac{\text{no. of operators}}{\text{shift}}\right) \times \left(\frac{\text{cost}}{\text{operators}}\right) \times \left(\text{annual demand}\right) \times \left(\frac{\text{cycle time}}{\text{unit}}\right) \Bigg/ \left(\begin{matrix}\text{existing}\\ \text{plant}\\ \text{capacity}\end{matrix}\right)$$

where the expected cycle time and the expected plant capacity are expressed as a fraction of an operator-shift-year (an operator-shift-year is equivalent to one operator for one shift for one year). The existing system can support only two shifts, and so an additional operation, consisting of two new presses, would be set up to meet excess demand. The expression for expected total costs for the FMS alternative would be similar to that given for CE above, and would include adjustments for maintenance costs and a portion of the capital cost. The advantages of the FMS over the existing system are a lower cycle time per unit and a smaller number of operators per shift. To illustrate the proposed methodology, we assume that the cycle times and plant capacity have uniform probability distributions. Moreover, suppose that the number of new products introduced in the future ($\Delta P = 0,1,2$) and the increase in annual demand for all products ($\Delta D = 0\%, -30\%, 30\%$) are random. For each state S_i, consisting of a pair $(\Delta P, \Delta D)$, $i=0,1,\ldots 8$, the mean of the total annual cost are calculated in Table 1 for the product and cost data of Table 2.

Clearly, excluding the apportioning of the capital costs to each year, the FMS, as expected, is preferred to the existing system based on the expected cost criterion. Thus, from an operational cost point of view, the existing system does not provide any "flexibility" relative to the FMS. However, this is not necessarily the case when two different FMS's are compared, and each one has its own advantages under some conditions. The measure defined in the next section is concerned with the evaluation of a

Table 1. FMS Example: Mean Total Annual Costs for Each State

State s_i	$(\Delta P, \Delta D)$	$Prob(s_i)$	FMS μ	Existing System μ
s_0	(0, 0)	0.02	467,057	494,384
s_1	(0,-30)	0.04	320,940	472,069
s_2	(0, 30)	0.04	603,574	831,700
s_3	(0, 0)	0.06	580,821	749,230
s_4	(1,-30)	0.12	406,575	485,086
s_5	(1, 30)	0.12	758,667	960,874
s_6	(2, 0)	0.12	699,385	951,576
s_7	(2,-30)	0.24	489,810	498,103
s_8	(2, 30)	0.24	910,161	985,049

Table 2. FMS Example: Product and Cost Data

Maximum number of shifts	2
Initial number of products	4
Initial demand	70,000 units per year per product
Cost of operator	$45,000 per year per operator
Cost of two small presses	$60,000
Capital cost of FMS	$1.9 million
FMS maintenance cost	$40,000 per year per FMS
Capital cost apportioned each year	20%
Cycle time of FMS	[35,50] uniformly distributed
Cycle time of existing system	[50,80] " "
Actual plant capacity	[1960,2040] " "

measure of "super flexibility" gained by a joint system that has the properties of both systems. This measure provides quantitative information on the disadvantage of the existing system.

5. DEVELOPING THE MEASURE

A description of the factors affecting the measure of "super flexibility" are given below.

Technology. By a technology $t \epsilon T$, we refer to any type of equipment or method that belongs to a production system. It could include management methods, such as group technology, as well as others.

Operational Problem. An operational problem $m \epsilon M$ is to be solved at the beginning of each planning period. For simplicity, it is assumed that each m is feasible for any $t \epsilon T$. We assume that the solution of the operational problem is determined at the beginning of this period for a given set of known and forecasted data. The known data corresponds to the system's characteristics, and the forecasted data is concerned with the state of the economy and, in particular, with demand (for simplicity, each planning period is assumed to be of equal duration). Let

p_m = The estimated probability that operational problem m will occur in a given time period.

Clearly, $\sum_{m=1}^{L} p_m = 1$ where $L = |M|$ (cardinality of M).

Problem Solutions. These feasible planning and control solutions $a_{mt} \epsilon A_{mt}$ are defined for operational problem m, given technology t. The a_{mt} could be identified in any data form. The set A_{mt} could be discrete or continuous and could be generated by any managerial method (for some examples, see Buzacott and Yao, 1986).

Other Factors. Let Y be a vector of random variables which may include the reliability, quality and processing time that affects the monetary value of a given problem. The randomness of these variables is due to the various disturbances, including machine errors and human errors, of a production system.

Using the above, we now define Z_t as the expected monetary value of a manufacturing system given that $t \epsilon T$ is implemented, an optimal solution $a^*_{mt} \epsilon A_{mt}$ is employed for a given m, and Y a vector of random variables which affects V, the random monetary value of t. By definition:

$$Z_t = E[V(a^*_{mt}, Y,m,t)] \tag{1}$$

By properties of conditional expectations, Z_t can be decomposed as:

$$Z_t = \sum_{m=1}^{L} P_m E[V(a^*_{mt} Y,m,t) | m] \ . \tag{2}$$

Letting $\mu(m,t)$ be the conditional expectation term in the above expression yields:

$$Z_t = \sum_{m=1}^{L} P_m \mu(m,t) \ . \tag{3}$$

Using a shorthand notation $\mu_t = \mu(m,t)$, we have

$$Z_t = E_m[\mu_t] \tag{4}$$

by definition.

A measure of the "super flexibility" for technologies t1 and t2, is given as follows:

$$F(t1,t2) = \sum_{m=1}^{L} P_m \max_{i=1,2} \mu(m,ti) - \max\{Z_{t1}, Z_{t2}\} \tag{5}$$

This monetary value measure evaluates the "flexibility" of a joint technology relative to the better of the technologies, given that they operate separately. It is important to note that F(t1,t2) may refer to an infeasible level of flexibility because of technological or physical constraints. Since F is a monetary measure, it evaluates the willingness of a risk-neutral decision maker to pay for the joint technology. Naturally, an FMS can be defined with reference to T as a technology t^* which satisfies the following condition:

$$F(t^*,t) = 0 \ \forall \ t \epsilon T. \tag{6}$$

Equation (6) holds if $\mu(m,t^*) \geq \mu(m,t)$ $\forall$ m and t, and by definition, $F(t,t) = 0$.

The above definition of an FMS provides a property of an economically optimal FMS with reference to a set of systems (technologies) and a monetary measure that takes into account only the operational aspects of the system. Following the definition above, a technology t is at least partially inflexible if there exists another technology $t2 \epsilon T$ such that $F(t1,t2) > 0$.

An important issue deals with a sensitivity analysis of $F(t1,t2)$ with respect to the distribution of μ_{t1} and μ_{t2}. To deal with (4) note that: $\text{Max} \{\mu_{t1},\mu_{t2}\} = \mu_{t1} + \text{Max} \{\mu_{t2} - \mu_{t1},0\}$.

Therefore, (4) can be rewritten as follows:

$$F(t_1,t_2) = E_m \ [\text{Max} \ \{\mu_{t2} - \mu_{t1},0\}] \geq 0 \tag{7}$$

given that, without loss of generality, $E_m(\mu_{t1}) \geq E_m(\mu_{t2})$.

6. ANALYSIS OF IMPLICATIONS

The analysis is based on the results of Mehrez and Stulman (1982) and Fatti, et al. (1987). For the convenience of the reader, the implications of their results will be reported here without reference to the original context, which deals with the economic value of perfect information (Mehrez and Stulman, 1982) and sample information (Fatti, et al., 1987) in a project selection problem.

a) For any underlying distribution function of $\mu_{t1} - \mu_{t2}$, a shift in $E_m(\mu_{t2})$ towards $E_m(\mu_{t1})$ will increase $F(t_1,t_2)$ (see for example Fatti, et al., 1987, p. 143). The implication is that the value of flexibility increases as the technological alternatives are "closer" in the mean sense.

b) For any underlying distribution of $\mu_{t1} - \mu_{t2}$, given that the $Var(\mu_{t1} - \mu_{t2}) = \sigma^2$, an upper bound on $F(t_1,t_2)$ is provided by $(1/2)\sigma^2$ (see for example Fatti, et al., 1987, p. 146). This upper bound is achieved by a symmetrical two-spiked bimodal distribution. This result can be used to establish a maximum price that a risk-neutral decision maker will pay under any circumstance for a "super flexible" joint technology. Since this price depends on the variance of $\mu_{t1} - \mu_{t2}$, the more negative the covariance between these two random variables, the higher will be the maximum price of "super flexibility." This is an intuitive result since a flexible technology performs better if it is an outcome of two negatively correlated technologies. It is important to note that the relationship between "super flexibility" and the standard deviation of $\mu_{t1} - \mu_{t2}$ is not necessarily linear, depending instead on the shape of its distribution.

c) For the class of symmetric and unimodal distributions the upper bound is lower $(\surd 3/4\sigma^2)$.

7. CONCLUSIONS

The purpose of this paper is to propose a statistical economic approach to the concept of flexibility. It is built on the concept of "super flexibility," which permits the computation of an upper bound on the maximum price an investor would be willing to pay for a joint system having the advantages of both given systems. Although this concept deals with only two systems, it can be extended to the case in which the managerial contribution of flexibility from a given system is evaluated with respect to the flexibility of several other systems, by employing an equation similar to (6). Actually, the previous results (a)-(c) also hold for this case.

Furthermore, the concept of the measure F could be extended to cover situations of a multi-period problem by employing traditional net present value (NPV) analysis and identifying a finite horizon framework based on a time index. The basic issue is concerned with the evaluation of the appropriate discount rate and the horizon. One alternative is to use the Capital Asset Pricing Model (CAPM), to estimate the discount rate under certainty (see for example Mullins, 1982). This multi-period case can address situations in which the flexibility of a system is enhanced over a series of learning periods.

The limitations of F arise from its inability to suggest a procedure that accomplishes the FMS selection decision. This measure at most guides the decision maker to look at possible marginal improvements of flexibility while few alternatives are given consideration. The mean criterion could lead to the optimal selection decision for FMS. Future research should focus on the engineering implications of the approach suggested here; in particular, on the advantages that different FMS's suggest under the various economic states. Such research may follow the lines pursued by Muramatsu, et al., (1985). Finally, we note that the sensitivity of the measure F to the shapes of the underlying distributions emphasizes the need for careful statistical economic evaluation of flexible manufacturing systems.

8. REFERENCES

Abernathy, W. J. (1978). **Productivity Dilemma: Roadblock to Innovation in the Automobile Industry**. Baltimore: Johns Hopkins University Press.

Ackoff, R. L. (1970). **A Concept of Corporate Planning**. New York: Wiley-Interscience.

Ackoff, R. L. (1981). **Creating the Corporate Future**. New York: John Wiley & Sons.

Arrow, K. J. (1951). **Social Choice and Multi Criteria Decision Making**. Cambridge: MIT Press.

Barad, M. and D. Sipper. (1988). "Flexibility in Manufacturing Systems: Definitions and Petri Net Modeling," **International Journal for Production Research, 26** (2), 237-248.

Bell, D. E. and P. H. Farquhar. (1986). "Perspectives on Utility Theory," **Operations Research, 36**, 179-183.

Brill, P. H. and M. Mandelbaum. (1989). "On Measures of Flexibility in Manufacturing Systems," **International Journal of Production Research, 27**, 747-756.

Buzacott, J. A. (1982). "The Fundamental Principles of Flexibility in Manufacturing Systems," **Proceedings of First International Conference in Flexible Manufacturing Systems**, 23-30.

Buzacott, J. A. and D. D. Yao. (1986). "Flexible Manufacturing Systems: A Review of Analytical Models," **Management Science, 32**, 890-905.

Canada, J. R. (1986). "Annotated Bibliography on Justification of Computer-Integrated Manufacturing System," **The Engineering Economist, 32**, 137-150.

Cunningham, A. A. and M. Mandelbaum. (1985). "Flexibility and Productivity in Manufacturing Systems," **Proceedings of IIE Conference**, Chicago, 404-413.

Demski, J. S. (1980). **Information Analysis**. Boston: Addison Wesley.

Fatti, L. P., A. Mehrez, and M. Pachter. (1987). "Bounds and Properties of the Expected Value of Sample Information for a Project-Selection Problem," **Naval Research Logistics Quarterly, 34**, 141-150.

Groover, M. P. and E. W. Zimmers. (1984). **CAD/CAM: Computer-Aided Design and Manufacturing**. Englewood Cliffs, NJ: Prentice-Hall.

Harrigan, K. R. (1985). **Strategic Flexibility: A Management Guide for Changing Times**. Boston: Lexington Books.

Hayes, R. H., S. C. Wheelwright and K. B. Clark. (1988). **Dynamic Manufacturing: Creating the Learning Organization**. New York: Free Press.

Ignizio, J. P. (1976). **Goal Programming and Extensions**. Lexington, MA: Heath.

Jaikumar, R. (1986). "Postindustrial Manufacturing," **Harvard Business Review**, 69-76.

Kapur, J. N., V. Jumar and O. Hawaleshka. (1985). "Maximum Entropy Principle and Flexible Manufacturing Systems," **Defense Science Journal, 35** (1), 11-18.

Keeney, R. L. and H. Raiffa. (1976). **Decisions with Multiple Objectives: Preference and Value Trade-Offs**. New York: Wiley.

Kochan, D. (1986). **CAM-Developments in Computer Integrated Manufacturing**. New York: Springer-Verlag.

Kumar, V. (1986). "On Measurement of Flexibility in Flexible Manufacturing Systems: An Information Theoretic Approach," **Proceedings of the Second ORSA/TIMS - Conference on Flexible Manufacturing Systems: OR Models and Applications**. K. E. Stecke and R. Suri, editors. Elsevier, New York, 131-142.

Kumar, V. (1987). "Entropic Measures on Manufacturing

Flexibility," **International Journal of Production Research, 25**, 957-966.

Kusiak, A. (1986). "Application of Operational Research Models and Techniques in Flexible Manufacturing Systems," **European Journal of Operational Research, 24**, 336-345.

Kusiak, A. (1987). "Artificial Intelligence and Operations Research in Flexible Manufacturing Systems," **Infor, 25**, 2-12.

Lazing, F. (1986). "Solving Multiple Criteria Problems by Interactive Decomposition," **Mathematical Programming, 35**, 335-361.

Lee, C. L. and E. V. Denardo. (1986). "Rolling Planning Horizons: Error Bounds for the Dynamic Lot Size Model," **Mathematics of Operations Research, 11,** 423-432.

Mandelbaum, M. and A. Buzacott. (1986). "Flexibility and Its Use: A Formal Decision Process and Manufacturing View." **Proceedings of the Second ORSA/TIMS Conference on Flexible Manufacturing Systems: OR Models and Applications**, K.E. Stecke and R. Suri, editors, Elsevier, New York, 119-130.

Marshak, T., and R.R. Nelson. (1986). "Flexibility, Uncertainty and Economic Theory," **Metroeconomica, 14**, 45-59.

Mehrez, A. and A. Stulman. (1982). "Some Aspects of the Distributional Properties of the Expected Value of Perfect Information (EVPI)," **Journal of the Operational Research Society, 33**, 827-936.

Mehrez, A. I., G. J. Krinsky, Miltenburg and B.L. Myers. (1987). "Flexible Manufacturing Systems: An Alternative Approach," Working Paper, Faculty of Business, McMaster University, Hamilton, Ontario, Canada.

Merkhofer, M. W. (1977). "The Value of Information Given Decision Flexibility," **Management Science, 23**, 716-728.

Monahan, G. E. and T. L. Smunt. (1989) "Optimal Acquisition of Automated Flexible Manufacturing Processes," **Operations Research, 37**, 288-300.

Mullins, D. (1982). "Does the Capital Asset Pricing Model Work?" **Harvard Business Review**, Jan.-Feb., 105-114.

Muramatsu, R., K. Ishii and K. Takahaski. (1985). "Some Ways to Increase Flexibility in Manufacturing Systems," **International Journal of Production Research, 23**, 691-703.

Nord, W. R. and S. Tucker. (1987). **Implementing Routine and Radical Innovations**. Lexington, MA: Lexington Books.

Pye, R. (1978). "A Formal Decision Theoretic Approach to Flexibility and Robustness," **Journal of the Operational Research Society, 29**, 215-227.

Rosenhead, J., and S. K. Gupta. (1968). "Robustness in Sequential Investment Decisions," **Management Science, 15** (2), B18-B29.

Rosenhead, J., M. Elton and S. K. Gupta. (1972). "Robustness and Optimality as Criteria for Strategic Decisions," **Operations Research Quarterly, 23**, 413-441.

Rosenthal, R. E. (1985). "Principles of Multi-Objective Optimization," **Management Science, 16**, 133-152.

Saaty, T. L. (1980). **The Analytic Hierarchy Process.** New York: McGraw Hill.

Sethi, S., and S. Chand. (1979). "Planning Heuristic Procedures for Replacement Models," **Management Science, 25**, 145-151.

Sink, D. S. and S. J. Devires. (1984). "An In-Depth Study and Review of State of the Art and Practice on Productivity Measurement Techniques," **Proceedings of May 1984 Annual Conference, Institute of Industrial Engineers**, 333-337.

Snader, K. R. (1986). "Flexible Manufacturing Systems: An Industry Overview," **Production and Inventory Management, 27**, 1-9.

Stecke, K. E. (1985). "Design, Planning, Scheduling, and Control Problems of Flexible Manufacturing Systems," **Annals of Operations Research, 3**, 51-60.

Stecke, K. E., D. Dubois, S. P. Sethi and K. Rathmill. (1985). "Classification of Flexible Manufacturing Systems: Evolutions Towards the Automated Factory," Working Paper No. 363, Graduate School of Business Administration, University of Michigan, Ann Arbor, USA.

Suri, R. (1985). "An Overview of Evaluative Models for Flexible Manufacturing Systems," **Annals of Operations Research, 3**, 61-69.

Wallsten, T. S. and D.V . Budesco. (1981). "Encoding Subjective Probabilities: A Psychological and Psychometric Review," **Management Science, 29**, 151-173.

Yao, D. D. (1985). "Material and Information Flow in Flexible Manufacturing Systems," **Material Flows, 2**, 143-149.

Zelenovic, D. M. (1982). "Flexibility as a Condition for Effective Production Systems," **International Journal of Production Research, 20**, 319-337.

LIST OF REFEREES

The Editor is particularly grateful to the many individuals who acted as referees for the papers in this volume.

Everett Adam, Jr.
Department of Management
University of Missouri
Columbia, MO 65211

Charles Alexander
College of Engineering
Temple University
Philadelphia, PA 19122

Joel H. Amernic
Faculty of Management
University of Toronto
Toronto, Canada M5S 1V4

Robert U. Ayres
Department of Engineering
Carnegie-Mellon University
Pittsburgh, PA 15123

R. Balachandra
Department of Management
Northeastern University
Boston, MA 02115

Gloria Barczak
Department of Marketing
Georgia State University
Atlanta, GA 30303

Carol A. Beatty
School of Business
Queen's University
Kingston, Canada K7L 3N6

Edward J. Blocher
Department of Accounting
University of No. Carolina
Chapel Hill, NC 27514

James Borden
Department of Accountancy
Villanova University
Villanova, PA 19085

P. H. Brill
Faculty of Business
University of Windsor
Windsor, Canada N9B 3P4

Jonathan Burton
Department of Management
Drexel University
Philadelphia, PA 19104

F. A. Bushman
School of Business
San Francisco State Univ.
San Francisco, CA 94132

Robin Cooper
Graduate School of Business
Harvard University
Boston, MA 02163

Arnould de Meyer
INSEAD
77305 Fountainbleau Cedex
France

Maurice Desplas
University Paris II
75270 Paris Cedex 06
France

Mohammad K. El-Najdawi
Department of Management
Villanova University
Villanova, PA 19085

John E. Ettlie
School of Business
University of Michigan
Ann Arbor, MI 48104

George F. Farris
Graduate School of Mgt.
Rutgers University
Newark, NJ 07102

Bela Gold
Drucker Graduate Mgt. Center
Claremont Graduate School
Claremont, CA 91711

Bruce Golden
College of Management
University of Maryland
College Park, MD 20742

Joel Goldhar
School of Business
Illinois Inst. of Tech.
Chicago, IL 60616

Thomas Gunn
Unisys Corporation
P.O. Box 500
Blue Bell, PA 19424

G. Hakala
Amerock Corporation
P.O. Box 7018
Rockford, IL 61125

James A. Hendricks
Department of Accountancy
Northern Illinois Univ.
Dekalb, IL 60155-2854

William A. Hetzner
Technology Assistance Program
Industrial Tech. Institute
Ann Arbor, MI 48106

Robert A. Howell
Howell Management Corp.
P.O. Box 106
Wilton, CT 06897

Mohan Kalkunte
AT&T Bell Laboratories
6200 E. Broad St.
Columbus, OH

Robert J. Kelsch
Xerox Corporation
800 Phillips Road
Webster, NY 14580

W. Carl Kester
Graduate School of Business
Harvard University
Boston, MA 02163

Cerry Klein
Dept. of Ind. Engineering
University of Missouri
Columbia, MO 65211

Paul Kleindorfer
The Wharton School
University of Pennsylvania
Philadelphia, PA 19104

G. Knolmayer
University of Berne
Hallerstrasse 6
CH-3012 Berne, Switzerland

Vinod Kumar
School of Business
Carleton University
Ottawa, Canada K1S 5B6

Andrew Kusiak
Dept. of Ind. & Mgt. Eng.
University of Iowa
Iowa City, IA 52242

James T. Mackey
Department of Accountancy
California State University
Sacramento, CA 95819

M. Mandelbaum
Atkinson College
York University
No. York, Canada M3J 1P3

Edwin Mansfield
Department of Economics
University of Pennsylvania
Philadelphia, PA 19104

Jack Meredith
College of Business
University of Cincinnati
Cincinnati, OH 45221

Wayne J. Morse
Dept. of Accounting/Law
Clarkson University
Potsdam, NY 13676

S. Y. Nof
School of Ind. Engineering
Purdue University
West Lafayette, IN 47907

O. Felix Offodile
Dept. of Admin. Sci.
Kent State University
Kent, OH 44242

Kay Poston
School of Accountancy
University of Missouri
Columbia, MO 65211

Everett M. Rogers
Dept. of Communications
Stanford University
Stanford, CA 94305

Ramesh Sharda
College of Business
Oklahoma State University
Stillwater, OK 74078

Steven Smith
Triplex Corporation
20316 Gramercy Place
Torrance, CA 90501

Luk N. Van Wassenhove
Katholieke Univ. Leuven
Leuven B-3030
Belgium 016 234931

C. A. Voss
University of Warwick
Coventry CV4 7AL
England

Donald J. Wait
Consultant-GE Co.
1247 Myron St.
Schenectady, NY 12309

Candace Yano
Dept. of Ind. & Oper. Eng.
University of Michigan
Ann Arbor, MI 48109-2117

Allen H. Seed
Arthur D. Little, Inc.
25 Acorn Pk.
Cambridge, MA 02140

C. Wickham Skinner
Department of Business
Harvard Business School
Boston, MA 02163

Thomas N. Tyson
School of Management
Clarkson University
Potsdam, NY 13676

Thomas E. Vollman
School of Management
Boston University
Boston, MA 02215

Samuel Wagner
Dept. of Business
Franklin & Marshall
Lancaster, PA 17604

Hans-Jurgen Warnecke
Fraunhofer-Institut
D-7000 Stuttgart
F. R. of Germany

MIX
Papier aus verantwortungsvollen Quellen
Paper from responsible sources
FSC® C105338

If you have any concerns about our products,
you can contact us on
ProductSafety@springernature.com

In case Publisher is established outside the EU,
the EU authorized representative is:
Springer Nature Customer Service Center GmbH
Europaplatz 3, 69115 Heidelberg, Germany

Printed by Libri Plureos GmbH
in Hamburg, Germany